THE GULF OF MEXICO
RESEARCH INITIATIVE

Carl A. Brasseaux and Donald W. Davis, Series Editors

THE GULF OF MEXICO RESEARCH INITIATIVE

A VISION FOR TWENTY-FIRST-CENTURY SCIENCE

JUSTIN SHAPIRO

Preface by Rita R. Colwell

University Press of Mississippi / Jackson

Publication of this book was made possible through funding provided by
the Gulf of Mexico Research Initiative.

The University Press of Mississippi is the scholarly publishing agency of
the Mississippi Institutions of Higher Learning: Alcorn State University,
Delta State University, Jackson State University, Mississippi State University,
Mississippi University for Women, Mississippi Valley State University,
University of Mississippi, and University of Southern Mississippi.

www.upress.state.ms.us

The University Press of Mississippi is a member
of the Association of University Presses.

Any discriminatory or derogatory language or hate speech regarding race,
ethnicity, religion, sex, gender, class, national origin, age, or disability
that has been retained or appears in elided form is in no way an
endorsement of the use of such language outside a scholarly context.

Copyright © 2025 by University Press of Mississippi
All rights reserved
Manufactured in the United States of America
∞

Publisher: University Press of Mississippi, Jackson, USA
Authorised GPSR Safety Representative: Easy Access System Europe -
Mustamäe tee 50, 10621 Tallinn, Estonia, *gpsr.requests@easproject.com*

Library of Congress Control Number: 2025022583

Hardback ISBN 9781496858719
Paperback ISBN 9781496858726
Epub single ISBN 9781496858733
Epub institutional ISBN 9781496858740
PDF single ISBN 9781496858757
PDF institutional ISBN 9781496858764

British Library Cataloging-in-Publication Data available

CONTENTS

LIST OF ACRONYMS vii

PREFACE BY RITA R. COLWELL. ix

INTRODUCTION: How Scientists Responded to a Major Oil Spill:
Building a Research Organization That Addresses a Societal Need 3

CHAPTER ONE: Deepwater Horizon and the Origin of
the Gulf of Mexico Research Initiative 13

CHAPTER TWO: Distributed Management:
Administering Scientific Research During an Unfolding Crisis 45

CHAPTER THREE: Open, Transparent, and Accessible Data:
GRIIDC and GoMRI Data Management Policy 75

CHAPTER FOUR: Outreach and Capacity: Building a Legacy
of Oil Spill Science in the Gulf of Mexico 93

CHAPTER FIVE: GoMRI Reflects: Synthesis and Legacy 113

CONCLUSION: The GoMRI Model: Social Considerations
and the Best Science 143

APPENDIX A: Gulf of Mexico Research Initiative
Research Board Members 157

APPENDIX B: Gulf of Mexico Research Initiative
Research Board Committees. 161

APPENDIX C: Gulf of Mexico Research Initiative Consortia 163

APPENDIX D: Gulf of Mexico Research Initiative
Management Team Members 165

APPENDIX E: Request for Proposal Timeline 169

APPENDIX F: Gulf of Mexico Research Initiative By-Laws 171

APPENDIX G: Gulf of Mexico Research Initiative Research
Board Code of Conduct. 175

NOTES 179

INDEX 199

LIST OF ACRONYMS*

BP: Formerly known as the British Petroleum Company, in 2001 it rebranded as BP. Hence, the name of the company is not officially an acronym.
AAAS: American Association for the Advancement of Science
AIP: American Institute of Physics
CSO: chief scientific officer
DARRP: Damage Assessment, Remediation, and Restoration Program
EPA: Environmental Protection Agency
GOMA: Gulf of Mexico Alliance, renamed Gulf of America Alliance in 2025
GoMRI: Gulf of Mexico Research Initiative
GRIIDC: Gulf of Mexico Research Initiative Information and Data Cooperative
LSU: Louisiana State University
MRA: Master Research Agreement
NAS: National Academy of Science
NGI: Northern Gulf Institute
NOAA: National Oceanic and Atmospheric Administration
NSB: National Science Board
NSF: National Science Foundation
PEMEX: Petróleos Mexicanos
RFP: request for proposals

RFQ: request for qualifications
RIS: Gulf of Mexico Research Initiative Research Information System
TAMU-CC: Texas A&M University-Corpus Christi

*The GoMRI consortia are generally referred to by their acronyms.
For a complete list of GoMRI consortia, see Appendix C.

PREFACE

MEETING THE NEEDS OF SOCIETY WITH RESEARCH

Rita R. Colwell, PhD

The Deepwater Horizon Oil Spill was tragic, with loss of lives and the resulting societal, economic, and ecological disaster. Despite government and industry preparations for oil spill response and the involvement of numerous expert groups, the amount of spilled oil and the geographic extent of the spill presented an extraordinary challenge. Much of the released oil was not recovered, dispersed, or burned.

An unusual decision was made by BP quickly after the massive spill occurred to fund scientific research relevant to the Deepwater Horizon spill that would provide science-based decision making for the way forward. The Gulf of Mexico Research Initiative (GoMRI) was established and challenged to provide new knowledge, tools, and solutions for prediction, prevention, mitigation, and remediation of the Gulf of Mexico disaster that would be innovative, effective, and useful for response to future oil spills. A total of $500 million was made available, with the challenge to accomplish research to address the fate and effect of the spilled oil and to advance underlying fundamental science. In accordance with its contractual agreement with

BP, the GoMRI Research Board commenced operations, making research awards and guiding GoMRI fully independently of BP. The ten-year duration of funding allowed research findings from the initial sample collections to inform follow-on research. The scale of work and the analyses and interpretation required understanding environmental drivers, notably those associated with seasonal and long-term resilience that could be harnessed for effective response.

The first step taken by the GoMRI executive team was to initiate sample collection: water, sediment, oil, biota, and all relevant samples that would be required for analysis, interpretation, and application. Management and design of the work were developed and deployed simultaneously. Various components of the research program are described in detail in the chapters of this book, specifically those areas of research and development addressing needs arising from the spill. Data gathered during the decade of research are preserved in the scientific literature and in the GRIIDC data depository (chapter three) and remain available to all users.

What is novel and highly relevant from the GoMRI program for science in the twenty-first century is that fundamental inquiry and innovative creativity were employed to address the very real and immediate needs of the communities of the Gulf of Mexico region who had been severely affected by the massive oil spill. Yet, the work that was done was both creative and functional. And as the work was being designed, implemented, and undertaken, the affected communities were fully informed on an almost daily basis—by public town hall meetings, community outreach, social media, and television. The research, as it progressed, was documented on film and the three documentary films are now historical depictions of the scientists at work.

The research accomplished during the ten years provided new discoveries that significantly advanced knowledge in fields of study relevant to response to oil spills and their effects. The many accomplishments included advanced molecular biological techniques that provided understanding of microbial degradation of spilled oil.

Other advances concerned the role of photochemical reactions in the fate of spilled oil. The interactions of microbiota in the water column, leading to formation of particulate matter that interacts with oil chemicals and results in transport to underlying seafloor sediment, were elucidated. The fate of spilled oil that came ashore on beaches and in marshes and physical dynamics in the water column that govern mixing of oil and water at small scales of centimeters to meters were expanded in understanding. The transport of oil at larger scales, from kilometers to hundreds of kilometers, and advanced modeling incorporating all of the preceding were developed. In addition to exploration of these phenomena, there were discoveries of new species of fish and other biota inhabiting the Gulf of Mexico.

The lasting message of the GoMRI program was the success of the mix of fundamental research and discovery, research focused on understanding the fates and effects of spilled oil, and practical options for response. This was accompanied by publication of results of the research in the peer-review literature and rapid and direct communication of discoveries and applications to the public and, most importantly, to users of the findings.

The partnerships that Gulf of Mexico–based scientists forged with colleagues in the Gulf region and elsewhere in the United States and internationally contributed to the success of GoMRI. Specific efforts to involve early-career scientists, students, and postdoctoral researchers ensured that all of the lessons learned will carry forward into the broader endeavors of the environmental sciences and science applied to national needs.

GoMRI provides a powerful model for interdisciplinary teamwork designed to put into action discoveries of research and serves as a paradigm for scientific research applied to national needs.

THE GULF OF MEXICO
RESEARCH INITIATIVE

INTRODUCTION

HOW SCIENTISTS RESPONDED TO A MAJOR OIL SPILL

Building a Research Organization
That Addresses a Societal Need

The Gulf of Mexico Research Initiative, a large-scale, industry-funded, multidisciplinary study of one of the largest maritime oil spills in history, represents a model for how the scientific community can organize its response to major disasters. The researchers that GoMRI funded focused specifically on the fate and effects of the oil that spilled into the waters of the Gulf, while also considering how spill responses might have affected the oil. Throughout most of the past century, scientists from a broad range of disciplinary backgrounds have done research with the objective of understanding the consequences of oil spills, especially in response to major spills. Those scientific efforts usually have been coordinated among scientists in colleges, universities, private research laboratories, and public institutions.[1] GoMRI represented a similar partnership between scientists, universities, and research institutions in response to a major oil spill, but with even closer coordination among research partners. It was quickly established during the first weeks and months following the start of the Deepwater Horizon disaster in April 2010. Its formation involved extensive negotiations and compromises between the federal government, the governments of the five states most affected by the spill, BP, and leading scientists and administrators.

Although it remains the largest marine oil spill by volume, Deepwater Horizon was not without precedent. The release of more than 130 million gallons of oil into the Gulf of Mexico by Deepwater Horizon was similar in scale to the 1979 Ixtoc I spill off the Yucatán peninsula in the Bay of Campeche.[2] As with the 1989 *Exxon Valdez* spill on the Alaskan coast, Deepwater Horizon attracted media attention throughout clean-up efforts. Observers of the *Exxon Valdez* and Deepwater Horizon spills watched as the oil slick slowly spread toward shorelines and first responders worked frantically to contain it in order to protect local ecosystems. Unlike the *Exxon Valdez* oil spill, however, in 2010 media outlets were able to broadcast live footage of the damaged wellhead and spreading oil slick for most of the duration of the spill. Viewers around the world could access live footage of the spill simply by turning on their televisions or going online.

Although Deepwater Horizon was preceded by several major oil spills that garnered national and international attention, the extent to which scientific research on the 2010 disaster was coordinated and funded during the weeks and months following the spill was noteworthy. GoMRI stands apart from previous scientific responses to major oil spills due to its funding source, degree of coordination among different universities and research institutions, its management structure, and its approach to scientific research.[3] Given that many scientists involved in GoMRI management and leadership had partaken in previous oil spill response efforts, GoMRI represents a refinement and improvement on what those scientists had learned. After a decade of operation, GoMRI proved successful, and serves as a useful model for how the scientific community can coordinate oil spill responses.

In 2010, as the general public and other stakeholders sought information about effects of the spill on their communities, the people involved in setting up the organization that became GoMRI began to assess the state of oil spill research capacity in the Gulf region. They adopted research program tools available at the time, then modified and adapted them in innovative ways to meet needs related to the Deepwater Horizon oil spill. During the following decade, GoMRI would build on extant experience and knowledge it had gained in laboratory experiments, experiments in the real world under controlled conditions, large-scale physics and engineering research facilities, field observations, and modeling at multiple scales, leaving new and improved scientific knowledge in its wake. The GoMRI Research Board—the topmost governance of GoMRI—fully recognized that its mission to fund the best possible oil spill science included education and outreach for those communities bordering the Gulf of Mexico. GoMRI prepared teaching materials, supported both undergraduate and graduate research, recruited scientists from an array of geographic locales and disciplines, and fostered a broad research program across five scientific themes, focusing on the research that would be of the greatest value to the Gulf region. GoMRI created and tested new opportunities to connect local communities to applicable scientific knowledge. Even though the primary mission of GoMRI was to fund the best science, the way that it pursued and connected science to relevant audiences for application represented an improvement in the scientific community's response to oil spills.

The notion of funding the best science—a foundational principle for GoMRI—was interpreted in a variety of ways over the course of the history of the initiative. Each chapter that follows highlights how GoMRI worked to connect the scientific research it funded to the needs of the larger Gulf of Mexico community as it recovered from Deepwater Horizon. In addition to supporting basic scientific research, the GoMRI mission included applying research results to meet the needs of the public, policymakers, and communities throughout the Gulf of Mexico. By 2015, the GoMRI Research Board began developing a comprehensive plan to synthesize the results of the research its funding had generated. By doing so, the GoMRI Research Board intended to broadly disseminate knowledge of the consequences of oil spills to relevant user communities and, thereby, inform any future spill responses. By the time funding allocated to research had been fully expended, the breadth of GoMRI had included research across five theme areas: data storage and management, outreach, education, capacity-building, and synthesis of existing information concerning oil spill science. It thereby established a legacy of improved detection, mitigation, and knowledge about oil spill effects.[4] The ways that GoMRI responded to conditions in the Gulf were modified as the research board received input from the scientific community, NGOs, and Gulf Coast residents. The GoMRI Research Board understood science had social implications, especially in the wake of a major disaster. Hence, GoMRI built its programs to facilitate translation of scientific knowledge generated to various audiences.

An aspect of GoMRI history that should be underscored is the willingness of its Research Board to adopt new ideas and programs. To support scientific research in the public interest, GoMRI was flexible and responsive to conditions in the field and among the various communities it served. Flexibility was an important part of the GoMRI operations and a necessity, considering the wish expressed by BP to establish the initiative as quickly as possible in 2010. The Research Board was fully open to improvement in management of its scientific, outreach, education, and capacity building programs.

The GoMRI mission clearly was to fund the best science, yet the initiative accomplished additional objectives, namely, to connect regional stakeholders and lay audiences to the scientific research being done on the oil spill in the Gulf of Mexico. The GoMRI Research Board understood well that scientific research alone would not suffice to mitigate the effects of the spill that had occurred. The research needed to be handed over to those communities in a way that they could put the new knowledge into practice. Broadly construed, this is, in fact, "science in the public interest" referenced throughout the following chapters. GoMRI developed programs to support the goal of using science to find ways to mitigate effects of oil spills. By supporting research on the consequences of Deepwater Horizon, the Research Board hoped the future capacity of the Gulf region to respond to spills and rebuild in the aftermath of disasters would be enhanced.

This book is structured thematically, rather than chronologically. Each chapter details a part of the GoMRI effort to support scientific research and disseminate knowledge to different communities. Many of the GoMRI objectives and goals were based on pursuing the public interest in different ways over time through outreach, education, and related efforts. "The public," of course, can mean many things and, where necessary, the different segments of the public that GoMRI targeted are described explicitly. In brief, the main observation presented in the narrative is that throughout its operational history, GoMRI employed a variety of programs and strategies to enhance the ability of the lay public to understand, respond to, and help mitigate the effects of an oil spill in the Gulf of Mexico.

The first of the following five chapters of the book, "Deepwater Horizon and the Origin of the Gulf of Mexico Research Initiative," contextualizes GoMRI's early history. It begins with a very brief description of the Deepwater Horizon platform explosion and subsequent spill. During the months following the blowout and explosion, various organizations and individuals played important roles in determining how the scientific community would respond to the disaster. Readers will be introduced to some of the major figures who were active in the weeks and months leading up to the formation of GoMRI, including Dr. Ellen Williams, the chief scientist for BP, and Dr. Rita Colwell, who would chair the GoMRI Research Board.

Chapter two, titled "Distributed Management: Administering Scientific Research During an Unfolding Crisis," describes the development of the GoMRI managerial structure, highlighting how the earliest days of the GoMRI operation required a great deal of flexibility on the part of those building GoMRI. During the four months that responders struggled to cap the well and slow the spread of oil toward the coastline, scientists were eager to get into the field and begin observing, monitoring, and recording data on the effects of the oil. The scale of the oil spill and the threat it posed to communities along the shoreline—not to mention the various industries that the Gulf supported—made Deepwater Horizon unpredictable in the short and long terms. Scientists knew they could act quickly to assist oil spill mitigation efforts and perhaps save the Gulf from a worst-case ecological disaster. Thus, the Research Board was encouraged to rapidly develop the scientific research program, but also incorporate new knowledge acquired during the decade it oversaw GoMRI.[5] Hence GoMRI employed a distributed management structure, encompassing many different organizations—all of which predated GoMRI—all organized along lines set forth by the Research Board. This meant that the board delegated day-to-day oversight of GoMRI operations to managers employed both by GoMRI and partner organizations.[6] Members of the Research Board relied on committees to plan new programs.[7] The committees reached out to partner organizations with established track records to provide the staff and logistical knowledge needed for a given effort. Colwell encouraged Research Board members to focus on those efforts deemed most important. The GoMRI leadership team, described in chapter two, maintained communication among the extensive network of organizations and their staff and managers, ensuring that researchers, outreach directors, educators, and other individuals working with GoMRI could progress toward their goals. GoMRI was a self-reflective organization, capable of flexibility adapting to changing circumstances during the hectic first months of the Deepwater Horizon disaster and as it generated new scientific knowledge during the following decade.

Chapter three, "Open, Transparent, and Accessible Data: GRIIDC and GoMRI Data Management Policy," provides a history of the GoMRI data management policy and how the Research Board held funding recipients to requirements for timely and accessible data submission. In 2010, as GoMRI was being organized, scientists in some fields were still accustomed to owning their own datasets as proprietary products. The Research Board, however, mandated that all recipients of funding make their data publicly available as soon as one year after its funding cycle concluded. GoMRI ended up storing about 98 percent of the data that its funding recipients generated, placing it into the Gulf of Mexico Research Initiative Information and Data Cooperative (GRIIDC) database. Chapter three concludes with a history of GRIIDC, which became the primary clearinghouse for all data and metadata generated with GoMRI funding. The initiative's commitment to producing open and transparent data ensures future generations of oil spill scientists and other interested parties will have access to a comprehensive database derived from the Deepwater Horizon spill.

Chapter four, "Outreach and Capacity: Building a Legacy of Oil Spill Science in the Gulf of Mexico," describes the GoMRI outreach, education, and capacity-building efforts. The chapter traces how those elements of the GoMRI scientific mission developed over time. The GoMRI outreach, education, and capacity-building programs extended the reach of scientific knowledge that funding recipients generated. The GoMRI Research Board instructed the consortia that GoMRI funded to find ways to bridge the gap between new scientific research results and the needs of the Gulf community. Internally, GoMRI also forged partnerships to assist in transmitting scientific findings to audiences across the United States and around the world. Thus, GoMRI was able to answer questions posed by the public about effects of previous oil spills and helped prepare the region in the event of any future disaster.

Chapter five, "GoMRI Reflects: Synthesis and Legacy," provides a history of the GoMRI synthesis and legacy efforts. GoMRI's approach to synthesis resulted in dozens of publications. By incorporating all scientific research about the role of oil in the Gulf of Mexico—not limited to research that it funded—GoMRI was able to better inform policymakers, first responders, local communities, and scientific researchers about oil in marine environments and the outcomes of different spill response strategies. Not unlike synthesis efforts in the wake of the *Exxon Valdez* disaster, the GoMRI Research Board, working with its consortia leaders, surveyed the state of knowledge at the time of the Deepwater Horizon disaster, understood how GoMRI-funded research improved the state of knowledge, and identified topics for further inquiry.[8] By doing so, GoMRI made scientific knowledge about oil spills in the Gulf of Mexico more accessible to other researchers, politicians, first responders, industry, and the general public. Readers with questions about the scientific knowledge that GoMRI funding generated will find answers in this chapter.

During 2010–2020, GoMRI funded a large number of scientific research projects concerning oil spills in the Gulf of Mexico. As of August 2024, GoMRI funding has resulted in 1,713 journal articles, 5,433 presentations, seven books, and 105 book chapters.[9] Scientists continue to draw on GoMRI-funded research, even several years after GoMRI ceased operation. Apart from the wealth of scientific publications, this book also provides a sense of how GoMRI engaged with the public, which is harder to quantify than the number of books, articles, or datasets supported by GoMRI. Research consortia receiving funding from GoMRI were required to do outreach to local communities on the Gulf Coast. Partnerships with organizations, including Sea Grant, allowed GoMRI to develop an understanding of concerns of the local communities and to connect those communities with information they needed. Synthesis of the new information enhanced understanding of oil spills by translating the new knowledge for user communities. GoMRI funded a series of documentaries that presented the fruits of its research to popular audiences. Research on the aftermath of Deepwater Horizon was not intended solely for delivery to scientific circles; rather, through partnerships with other organizations and the consortia, GoMRI strove to connect communities around the Gulf of Mexico to the new knowledge that its funding had generated. This book describes, in part, those efforts to build connections between scientific experts and local communities. To do so, it is first necessary to describe the context in which GoMRI emerged and how it was organized to facilitate such a large-scale scientific program.

CHAPTER ONE

DEEPWATER HORIZON AND THE ORIGIN OF THE GULF OF MEXICO RESEARCH INITIATIVE

As environmental challenges become increasingly more complex, so too will be anticipated necessary societal responses. Success will be achieved if support of the entire community is gained, including extant partnerships, collaborations, and organizations from all sectors, government, academia, commerce, and non-profit, employing the most advanced knowledge, especially that gleaned from fundamental research. GoMRI was a proverbial "airplane constructed while flown" organization, yet it succeeded in employing the best, most effective science to address and remediate an ongoing environmental disaster, while providing lessons applicable to future environmental catastrophes, especially those anticipated in a changing global climate.
—DR. RITA COLWELL, GoMRI Research Board chair

On April 20, 2010, eleven workers lost their lives when an explosion occurred aboard the Deepwater Horizon offshore drilling platform.[1] Survivors on the rig dashed across the platform and floors underneath, searching for refuge from the fire and smoke. Seeking lifeboats to take them to safety, crew members rushed to escape, "certain they were about to be cooked alive, scrambl[ing] into enclosed lifeboats for shelter, only to find them like smoke-filled ovens."[2] Meanwhile, high above on the bridge, the platform control engineers felt severe shaking below. Computer screens showed something was terribly wrong, though from their perch high above the engineers were unaware of the scene on the platform. Andrea Fleytas, a twenty-three-year-old technician who had been on the rig for eighteen months, began responding to sensors and alarms reporting that the situation on the floor was increasingly dire. She quickly decided to press the distress button that summoned the Coast Guard, which evacuated survivors from Deepwater Horizon as it began to collapse into the Gulf.[3] Five thousand feet below the crumbling platform, the broken wellhead began spewing oil at a rate of over 50,000 barrels per day. It would continue to do so for nearly five months, until September 19, 2010.

At BP headquarters in London, news of the explosion began circulating in the early morning hours. The evening news cycle in the United States was already winding down during the first minutes and hours after the explosion, when most viewers in the United Kingdom were already asleep. Dr. Ellen Williams, who had been serving as the company's chief scientific officer for just four months, recalled walking into the office the following day to find her colleagues in shock. Although BP was aware that an explosion had occurred at one of its drilling sites, few knew in those first hours what had caused the explosion or how dire the situation was.[4]

While some of the company's executives headed to New Orleans to assess damage and monitor the spill, others remained in London to figure out BP's response strategy. The company hoped to maintain a degree of trust among local, state, and federal politicians in the days and months that followed. Although federal courts would eventually find that BP was primarily responsible for the disaster, during the early weeks of Deepwater Horizon, most of the legal proceedings took place behind the scenes, allowing BP to mount an aggressive public relations campaign. Audiences across the country watched as

Figure 1.1. The US Coast Guard searches for survivors of the explosion on the Deepwater Horizon platform as first responders battle the flames. Science History Images / Alamy Stock Photo.

NOAA Office of Response and Restoration team members, BP, and the Coast Guard struggled to limit the flow of oil, broadcasted live on cable news and streamed online. As the initial response continued, BP senior executives considered contributing $500 million for research on oil spills in the Gulf of Mexico. Robert Dudley, who at the time served on the BP board of directors as managing director of activities in Asia and the Americas, handed the idea to Williams, who was put in charge of determining who would be responsible for developing a $500 million research program.[5] Dudley was quickly summoned to the Gulf to oversee the company's response to Deepwater Horizon. He would eventually replace the beleaguered CEO, Tony Hayward, after a disastrous public relations campaign and his subsequent resignation announcement on July 27, 2010.[6]

Dudley recognized that scientists could play an important role in the response to Deepwater Horizon. He believed that funding scientific research would assist BP with its public relations effort during the spill. Announcing that the company was offering half a billion dollars to research the spill might reduce hostility toward BP, at least from policymakers who would be determining how to punish the responsible parties. Dudley also understood it was necessary to get scientific researchers to work in the field as soon as possible. Researchers across the United States were eager to launch data collection and monitoring programs to understand the scale and scope of the spill. Some organizations had made money available quickly to researchers. The National Science Foundation provided dozens of RAPID grants to move researchers into the field.[7] The National Oceanic and Atmospheric Administration's (NOAA) Damage Assessment, Remediation, and Restoration Program (DARRP) was also initiated in the first weeks following the start of the spill. DARRP and GoMRI would have some overlapping research responsibilities, which, according to Research Board member Dr. Peter Brewer, created some initial tension between the two research entities. Still, as Brewer maintains, both sides managed their efforts professionally and with courtesy. While GoMRI was able to share all of its data, DARRP had to cautiously protect its data as the process for assigning damages for the spill continued in federal courts.[8] Through DARRP, NOAA was able to begin evaluating "the type and amount of restoration needed in order to return the Gulf to the condition it would have been in before the spill and to compensate the public for the natural resource services that were injured or lost."[9]

On May 19, as BP was in the initial stages of figuring out how to distribute half of a billion dollars in research funding, the federal government invited the leadership of its various environmental, resource management, and disaster response agencies to meet with academic researchers at the headquarters of the Environmental Protection Agency. Among those asked to attend the conference were Ken Salazar, then Secretary of the Interior; EPA Administrator Lisa Jackson; Dr. John Holdren, director of the Office of Science and

Technology Policy; Commandant of the Coast Guard Thad Allen; and Dr. Marcia McNutt, director of the United States Geological Survey. The purpose of the meeting was to "refine the short- and long-term oil spill research needs and identify the reserves and assets within the scientific community that can be utilized and integrated into our national response and mitigation strategies."[10] The federal government wanted to coordinate with academic and research institutions to develop a scientific program for investigating the Deepwater Horizon oil spill.

The research framework that came out of that meeting was similar to the one that GoMRI would later employ. Some of the short-term research goals that federal officials and oil spill scientists identified included understanding undersea plumes, mid-water column ecosystem impacts, and the question of how hurricane season—just one month away at the time of the meeting—could affect the fate of the oil. The long-term research agenda called for investigations into the role of dispersants in the marine ecosystem, the best approaches for monitoring deep-water ecosystems, and how scientists could learn from natural oil seeps. Many additional research goals were listed in the invitation.[11] What the meeting ultimately produced was a set of questions that researchers and their federal partners believed were the most important ones that scientists could address. This was an early step toward shaping the scientific response to Deepwater Horizon.

Five days after the meeting at the EPA, BP publicly announced its intention to provide $500 million to support scientific research into the impact of the oil spill on the Gulf environment. The company's goal was to build a framework for an effective research initiative that could fund and administer interdisciplinary research teams. Rather than relying on several different academic and research institutions, the company wanted to build a single organization coordinating much of the scientific response. To begin developing an organization capable of overseeing a large-scale scientific program, BP asked Williams to help set up an independent advisory committee. This committee was to be comprised of leading scientists and administrators in fields pertinent to oil spill science and the effect of oil on the marine environment. Members of the independent advisory committee would later serve on the GoMRI Research Board. The committee was tasked with developing research policies and priorities as well as a plan to spend BP's funds appropriately.

While Williams began searching for members of the independent advisory committee, BP contacted Louisiana State University to ask for its support in the formation of the burgeoning scientific organization. In return for its help, BP offered the university a grant to kick-start scientific research in the Gulf as the spill continued. LSU had a long-standing relationship with BP. According to a BP press release that detailed the negotiation, "LSU has a significant amount of experience in dealing with the oil and gas industry and deep knowledge pertaining to the Gulf of Mexico across numerous topical disciplines."[12] A partnership with LSU would offer several advantages. For example, LSU could help coordinate a broad interdisciplinary approach to research on the oil spill. Its faculty and administration included some of the most experienced marine biologists, oceanographers, physicists, and research administrators in the country. Many individuals who would play key roles in GoMRI had professional connections to LSU.

Apart from LSU, BP had other institutional connections that would prove advantageous to formation of the Gulf of Mexico Research Initiative. The company had collaborated with the Scripps Institution of Oceanography since 2004 and Texas A&M University in Galveston, the Monterey Bay Aquarium Research Institute, the University of Aberdeen, the National Oceanography Centre in Southampton, UK, and the University of Glasgow since 2008. Very few institutions of higher education or research in the Gulf of Mexico region lacked ties to the oil and gas industry. Those ties might be very direct, as when oil companies offered research funding to colleges and universities. They might be indirect, as when scientists left posts at academic institutions to work for one of many oil and gas companies in the region, or vice versa. This situation required the independent advisory committee to be very careful about who it picked to join GoMRI. From the outset, its members scrupulously avoided recruiting individuals who might be inclined to sacrifice the best possible relevant science out of a professional or financial commitment to the oil and gas industry. Science is, after all, a social activity. Hence, it is always subject to social pressures. Yet it is possible to conduct scientific research (and distribute funding for such research) in fair and ethical—if not perfectly neutral—ways. Members of the GoMRI Research Board were acutely aware of the scrutiny their organization would undergo by the press, the state and federal governments, and the public. As discussed in the following chapter, its members established rigorous codes of conduct and ethics to ensure the full board's impartiality in overseeing the GoMRI scientific program.

As BP began calling on academic institutions to help support research on Deepwater Horizon, Williams was close to choosing someone who she considered to be an ideal candidate to lead the independent advisory committee. She reached out to Dr. Rita Colwell, a distinguished environmental microbiologist who was then a faculty member of the University of Maryland–College Park. Her research included studies on oil in the environment, notably microbial degradation and ecology of spilled oil. Over the course of her career, Colwell had built an impressive CV and a towering reputation in the scientific community. In 1998 President Bill Clinton selected her to serve as the first female director of the National Science Foundation, which she would lead until the expiration of her term in 2004. While serving as director, Colwell oversaw all NSF activities, from "the development of policy priorities to the establishment of administrative and management guidelines."[13] Congress established the National Science Foundation in 1950 to promote scientific progress, set research priorities for the federal government, and to benefit the public and national defense through scientific research. It has played a crucial role in supporting new developments in both science and technology throughout the Cold War and into the twenty-first century. During her term as director of the NSF, Colwell focused on national priorities such as K–12 STEM education, graduate science and engineering education, and increasing the number of women and people of color engaged in scientific research and engineering.

In 1977, prior to her work for the NSF, Colwell founded the University of Maryland–College Park Sea Grant program. Sea Grant is a partnership between the federal government (predominantly the National Oceanographic and Atmospheric Administration) "and thirty-four university-based programs in every coastal and Great Lakes state, Puerto Rico, and Guam."[14] It draws on the knowledge of thousands of scientists, engineers, public outreach experts, and others to help citizens understand, protect, and preserve coastal regions in the United States. As described in later chapters, Sea Grant was a vital partner in GoMRI's outreach efforts. Many individuals, including Colwell, had worked for Sea Grant in the past, including GoMRI's Chief Scientific Officer Dr. Charles "Chuck" Wilson. During her time with Sea Grant, Colwell became familiar with nonscientific communities and how they learned about the state of the coasts on which they depended. Her time at Sea Grant would prove useful as GoMRI developed.[15]

The path that Colwell's professional career took meant that she worked closely with—and frequently led—federal agencies and national scientific organizations. Her familiarity with standards and best practices in scientific administration suited her well in her new leadership role for what would become the GoMRI Research Board. In addition to her term as the director of the NSF, she had previously served on the National Science Board (NSB) from 1984 until 1990. The NSB works with the director of the NSF to "recommend and encourage the pursuit of national policies for the promotion of research and education in science and engineering."[16] This means two things. First, the NSB establishes National Science Foundation policies within the framework of larger pieces of legislation put forward by Congress or the science and technology goals of a presidential administration. Second, the NSB acts as an advisory body to the President and the Congress "on policy matters related to science and engineering and education in science and engineering," occasionally issuing policy papers on pressing scientific matters.[17] Thus Colwell had experience both as a scientific advisor and as administrator of a major federal research organization, making her an excellent candidate to lead GoMRI.

Colwell also had an impressive career outside of her work for the federal government. Prior to 1998 she was the president of the University of Maryland Biotechnology Institute and Professor of Microbiology and Biotechnology at the University of Maryland, College Park. Colwell had also received national and international renown for outstanding work in her field. At different times, she had served as president of the American Society of Microbiologists; president of the American Institute of Biological Sciences (AIBS), which would later become a partner to GoMRI; director of the University of Maryland Environmental Sciences Program; and a number of other prestigious positions in her field of study. Given her credentials, her experience crafting and executing policy for the National Science Foundation, and her lack of any prior professional connection to BP, Colwell was the favored candidate to build the organization that would become the Gulf of Mexico Research Initiative. She had allies in industry, the federal government, and the scientific community and was a widely respected scientific authority.

Colwell and Williams also shared an institutional connection. Although at the time of the spill Williams was the chief scientific officer for BP, she had previously been a member of the faculty of the University of Maryland, College Park. Colwell has been a tenured professor at the University of Maryland since 1972 and performed most of her research and teaching just a few buildings away from where Williams herself worked until joining BP in January 2010. According to Williams, although the two did not work closely together during their time at the University of Maryland—they were in different departments—the personal connection they had as members of the same university faculty made Williams's call to Colwell easier.[18] After consulting with her colleagues, Williams phoned Colwell to gauge her interest. On May 26, 2010, two days after BP announced its $500 million grant, Williams formally asked Colwell to head up the independent advisory committee—which will subsequently be referred to as the Research Board—for the Gulf of Mexico Research Initiative.[19]

Colwell immediately took on a flurry of work. Williams and other BP executives hoped to approve the first research grants by the end of 2010. That was no easy task. The GoMRI Research Board would have to figure out how to manage expenses for the board and a fledgling staff, develop requests for proposals (RFPs), establish a proposal review process, determine in a very short period of time how the grants would be awarded and managed, and where and how to store the data collected. Furthermore, the board needed to figure out how to maintain independence from BP through robust conflict-of-interest policies. It also had to develop a comprehensive scientific program to study and monitor both the immediate crisis at hand and the long-term consequences of the Deepwater Horizon spill. Using the NSF as a model would help with large-scale and longer-term projects, but the immediate problem the board encountered was how to support scientists eager to start field work as soon as possible.

The newly appointed members of the Research Board faced many challenges, took their work seriously, and moved at a rapid clip. By June 12, the board comprised five members, in addition to the chair, Rita Colwell. All of the inaugural board members—Dr. Margaret Leinen, Florida Atlantic University Harbor Branch; Dr. David Halpern, NASA Jet Propulsion Laboratory; Professor John Shepherd, National Oceanography Centre, University of Southampton, UK; Dr. Jürgen Rullkötter, University of Oldenburg, Germany; and Dr. Jorg Imberger, University of Western Australia—were selected in discussions between Williams and Colwell. Those individuals comprised the initial board and would later serve on GoMRI's twenty-member Research Board.[20] The diverse backgrounds and expertise of the initial six-member Research Board represented GoMRI's broad and interdisciplinary approach to research on the Gulf of Mexico.[21]

Meanwhile, BP was aware that scientists hoped to start conducting field research quickly. On June 15, 2010, during a meeting at LSU that was sponsored in part by NOAA, BP announced the pending release of $45 million in funding in order to get researchers into the field to perform baseline studies and begin collecting oil sample data.[22] This initial $45 million "jump start" was not under the purview of the Research Board but was included as part of the $500 million commitment from BP. These initial funds were distributed to several Gulf state institutions and programs prior to Research Board activity. With oil continuing to flow from the damaged wellhead on the floor of the Gulf, researchers were eager to begin gathering data in order to develop mitigation strategies, among other pressing issues.

$500 million is quite a lot of money, and BP and the Research Board were not the only parties interested in how it would be spent. Given that their coastal communities would bear the brunt of the disaster, Gulf state governors believed that the states should play a primary role in deciding how the research funds were to be distributed. On June 16, the day after BP announced the release of $45 million in rapid response funding, the White House froze funding for GoMRI until all parties could reach a consensus on how to build the GoMRI program.[23] Over the course of several months and into autumn of 2010, the governors negotiated with BP and the initial Research Board members to ensure that the funding would be spent predominantly in the five Gulf states. From the board's perspective, the concern about such a limitation was that by doing so, the money might not be spent on the most comprehensive science, but rather on local talents and priorities. BP and the board sought an open and competitive scientific program, encompassing the best institutions and universities from around the world. Furthermore, some Gulf state institutions had longstanding and close financial relationships with major oil and gas producers. The Research Board wanted to ensure that political influence and prior industry ties would play no role in determining where BP's financial commitment would be spent. The next two sections describe how GoMRI established its independence from political influences.

GOMA to GoMRI: Development of GoMRI Governance

The Gulf state governors exerted most of their influence on the development of GoMRI through the Gulf of Mexico Alliance (GOMA). Brought on as a party to GoMRI in 2010, GOMA soon came to play a significant logistical and financial role in GoMRI operations.[24] The history of the Gulf of Mexico Alliance stretched back a decade to 2000. In August of that year, Congress passed the Oceans Act of 2000, which provided federal funding for the first comprehensive review of the oceans, coasts, and Great Lakes of the United States undertaken in thirty-five years.[25] The Oceans Act also created the United States Commission on Ocean Policy, which consisted of seventeen individuals from a wide range of positions in the public and private sectors as well as leading research institutions.[26] Congress tasked the commission with eight broad goals, including stewardship, promoting research and knowledge about the oceans and coasts, protecting life and property against both natural and human-caused hazards, and enhancement of transportation and commerce in marine environments.[27]

In its final report, the commission issued policy recommendations on a wide range of topics, following the mandate that Congress initially set forth in 2000. Titled "An Ocean Blueprint for the 21st Century," the Commission's report issued one recommendation that would spur creation of GOMA. Chapter five detailed the need for a regional approach to ocean research and ecosystem management.

"The voluntary establishment of regional ocean councils," the report stated, "... would facilitate the development of regional goals and priorities and improve responses to regional issues."[28] The commission recommended the formation of five such regional ocean councils, including the Chesapeake Bay Program, the Delaware River Basin Commission, the California Bay–Delta Authority, the Great Lakes Region council, and the Gulf of Mexico Program.[29]

The Council's recommendation led to the formation of the Gulf of Mexico Alliance in 2004.[30] By design, GOMA would work with Gulf state and Federal partners in the surrounding community to initially develop and later update action plans for the Gulf state governors. The action plans reflected different regional priorities at different times; as of 2024 there have been four action plans (2004–2009, 2009–2015, 2016–2021, and 2021–2026). For example, from 2016 until 2021, GOMA's priorities included Coastal Resilience, Data and Monitoring, Education and Engagement, Habitat Resources, Water Resources, Wildlife and Fisheries, Marine Debris, and Ecosystem Services. In order to research and prepare publications and strategies addressing the region's priorities, GOMA enlisted support from a number of different partner agencies at both the state and federal levels.[31] The 2016 action plan was written with help from the Florida Department of Environmental Protection, which coordinated the Data and Monitoring Team; the Mississippi Department of Environmental Quality, which led the Water Resources Team; and the Harte Research Institute, housed on the campus of Texas A&M University-Corpus Christi (TAMU-CC), which organized the Wildlife and Fisheries Team.[32] Each state was also made responsible for at least one of the priorities that GOMA adopted during its formation. The federal government was represented within GOMA by the federal workgroup, a collaborative effort among thirteen federal agencies including the Environmental Protection Agency, the National Oceanic and Atmospheric Administration, the US Fish and Wildlife Service, and others.[33] The federal workgroup assisted GOMA in carrying out its responses regarding the regional issues that it uncovered. Collaboration among these different groups accomplished several crucial projects, such as a warning system to forecast the occurrence of algal blooms and the identification of potential future dredging projects to restore marine habitats across the region.[34]

The Gulf of Mexico Alliance served as a model for the ways that diverse state, federal, and private actors could be brought together to attend to basic scientific inquiry, restoration, and research in the Gulf of Mexico region. GOMA coordinated research on the Gulf Coast ecosystem and came up with ways to address and mitigate some of the environmental degradation that had occurred over the course of the nineteenth and twentieth centuries. Within the Gulf of Mexico Alliance's network of partnerships, federal and state agencies could work together to address some of the most pressing problems facing the Gulf ecosystem. Those partnerships, along with GOMA's history of supporting research and remediation around the Gulf Coast, would prove very helpful to GoMRI.

In 2010, GOMA came to play a major role in administering the Gulf of Mexico Research Initiative. Once BP decided to disburse the $500 million in funding, the Research Board and BP were faced with the question of how to build a credible and transparent financial apparatus to handle such a large sum of money. Ultimately, BP selected the Gulf of Mexico Alliance to manage disbursement to the Gulf of Mexico Research Initiative. Most aspects of financial support for the scientific program of the Gulf of Mexico Research Initiative came under the authority of GOMA.[35] In addition, BP worked with GOMA to establish a three-way agreement with Consortium for Ocean Leadership, which had an established capacity for administering large research contracts, to share in some of the contractual duties. Therefore, some financial aspects of GoMRI came under the authority of other nonprofit organizations. Funds for the Research Board were managed by the American Institute of Biological Sciences (AIBS), as discussed below. GOMA and Ocean Leadership were responsible for managing research contracts and support staff for the bulk of the $500 million contribution, overseeing its disbursement to scientific researchers and contracted partners that were awarded GoMRI funding.

With the Gulf of Mexico Alliance in place as the financial apparatus for the Gulf research program, GOMA, after consulting with the Research Board, began hiring individuals with experience administering large scientific endeavors. In January 2011, Dr. Michael Carron was brought on as the program director of the Gulf of Mexico Research Initiative, having been a key consultant to the Research Board in the months leading up to his official appointment. Before moving to GoMRI, Carron had served as director of the Northern Gulf Institute (NGI), which had been involved with the first allocation of $45 million from BP in June 2010. Carron contacted Kevin Shaw, who would eventually work as the program manager for GoMRI, in the fall of 2010 to help draft the Master Research Agreement and oversee the flow of funding throughout the nascent Gulf of Mexico Research Initiative.[36] The Master Research Agreement was the result of negotiations between BP and the Gulf of Mexico Alliance (GOMA) about how to administer and manage the $500 million allocation. It served as a constitution for GoMRI, outlining its mission and providing a framework for its scientific program.

When BP and GOMA finally approved the Master Research Agreement on March 14, 2011, eleven months after the beginning of the Deepwater Horizon disaster, the document limited BP's role in operations and design of the Gulf of Mexico Research Initiative. According to the Master Research Agreement, members of the Research Board must "have peer-recognized research credentials and be from academic institutions, or been associated for long periods with academic institutions, or from other nationally-recognized research entities such as a national laboratory, research institute, or other peer-recognized research entity."[37] Furthermore, "appointees shall not include political appointees, BP employees, or State personnel outside of academic or research institutions."[38] The inclusion of such statements pertaining to the professional credentials of Research Board appointees deftly anticipated potential concerns about actual or perceived political and financial conflicts of interest. These stipulations ensured that members of the Research Board would be qualified to carry out their responsibility to GoMRI and would not be beholden to their employers in their decision-making. Although the Master Research Agreement did give BP the option to object to appointments, which would trigger mediation between GOMA and the company, it never exercised that right.

After negotiating with BP during the fall of 2010, the Gulf state governors secured two appointments each to the GoMRI Research Board. Subsequently, BP asked the Research Board to identify four additional at-large members to arrive at a total of twenty members. Colwell and the earliest members of the Research Board took steps to ensure that scientific integrity and expertise would motivate the decisions made at the board level, including providing a comprehensive list of outstanding Gulf of Mexico research scientists to the governors and BP as potential nominees to the Research Board. BP, the Research Board, and the governors reviewed the nominees and agreed to their appointments without dispute.[39] As is customary with corporate boards, BP specified an annual honorarium for each Research Board member.

Rather than immediately discussing how to spend the $500 million, the earliest Research Board meetings focused on questions about developing a framework for equitable and fair allocation of funding that would best support the GoMRI mission. According to Margaret Leinen, who would serve as the board's vice chair, the first board meetings were about three topics: governance, the GoMRI scientific concepts, and implementation of the scientific program. Conversations about governance pertained to the way that GoMRI would be administered and managed. Conversations about scientific concepts were about the main research subjects that GoMRI would support. The result of conversations about scientific concepts were the five research themes that GoMRI would pursue:

1. Physical distribution, dispersion, and dilution of petroleum (oil and gas), its constituents, and associated contaminants (such as dispersants) under the action of physical oceanographic processes, air-sea interactions, and tropical storms.
2. Chemical evolution and biological degradation of the petroleum/dispersant systems and subsequent interaction with coastal, open-ocean, and deep-water ecosystems.
3. Environmental effects of the petroleum/dispersant system on the sea floor, water column, coastal waters, beach sediments, wetlands, marshes, and organisms; and the science of ecosystem recovery.
4. Technology developments for improved response, mitigation, detection, characterization, and remediation associated with oil spills and gas releases.
5. Impact of oil spills on public health including behavioral, socioeconomic, environmental risk assessment, community capacity, and other population health considerations and issues.

Conversations about implementation of the scientific program pertained to evaluation and selection of research proposals. This process is described in chapter two. By focusing on the larger questions about what GoMRI should accomplish and how it could best carry out its mission, Research Board members were able to agree on a set of shared values. This contributed to the collegial relationship among board members.[40]

Soon after the Master Research Agreement was finalized and the Research Board was in place, it became clear to Colwell and other members of the board that there was a pressing need for a chief scientific officer. The board felt that it needed to give an individual the ability to oversee and streamline both the scientific and administrative work done at GoMRI's highest levels. Leinen stated in an interview that the chief scientific officer would have been helpful even earlier in GoMRI's existence. Prior to the appointment of a CSO, the board was dealing with day-to-day problems while also discussing larger questions of implementation and governance that would guide GoMRI during the rest of the decade.[41] To take some of the burden off the Research Board, the chief scientific officer would be responsible for providing "scientific and research advice and leadership to . . . GoMRI," coordinating between the Research Board and various partners working with GoMRI, and acting as liaison to GOMA and BP.[42] A Research Board committee chaired by Chuck Wilson was formed in 2011 to identify suitable candidates for the position of chief scientific officer (CSO). Shortly after the committee was established, Wilson discussed the requirements of the position with Colwell and his interest in it. Considering his capabilities and experience, Colwell suggested he consider being a candidate. To do so, Wilson resigned from the committee and Colwell named a new chair of the selection committee. After an open search, followed by in person interviews, the committee determined that Wilson, based on his knowledge of the Gulf of Mexico, expertise in the marine sciences, and experience as an administrator, would be the most suitable candidate to serve as CSO.

Wilson had initially joined the Gulf of Mexico Research Initiative in December 2011 as one of Louisiana Governor Bobby Jindal's appointees to the Research Board. Wilson had served as LSU Vice Provost for Academic Affairs, as well as chair of the Department of Oceanography and Coastal Sciences and executive director of the Louisiana Sea Grant College Program at the same institution. In those roles, Wilson developed and followed institutional guidelines for research funding. His experience handling academic grants and reviewing applications for research and travel funds, among other administrative responsibilities, translated well to a large organization like the Gulf of Mexico Research Initiative. Wilson had also known Colwell for several decades. The two first met at an annual meeting of the American Society of Microbiology when Colwell was in her early career and Wilson was a graduate student. Since that initial meeting, they had both advanced in their fields with distinction. Colwell appreciated that Wilson had a kind and jovial demeanor when carrying out his professional responsibilities.[43] She believed he would be a perfect fit for the role of chief scientific officer, especially given his familiarity with the Gulf Coast academic community.

After Wilson had been appointed chief scientific officer for GoMRI on February 24, 2012, the Research Board greatly expanded the responsibilities of his position. Although he was no longer a voting member of the Research Board, he attended meetings as an *ex officio* member. The CSO served as liaison between scientists who received GoMRI funding and the Research Board. Wilson helped develop and improve the RFP process. He also had oversight over data submission, ensuring that GoMRI-funded recipients adhered to the data management stipulations. In addition to day-to-day responsibilities, Wilson also played an important role in connecting GoMRI with outside parties. He served as a liaison to Colwell, the Research Board, and BP. As the designated liaison between BP and GoMRI, Wilson and Colwell conferenced almost daily about GoMRI matters. Colwell was required to respond quickly to questions from the media about research in the Gulf and Wilson worked with her to draft public statements and press releases.[44] Wilson represented GoMRI to local, state, and federal agencies. His mandate was broad and important in maintaining internal communication and the scientific independence of the organization.

Building Ethical Scientific Research

During GoMRI's formative years, BP found itself in an unfamiliar position. The company had a long tradition of sponsoring in-house scientific research, but in the case of GoMRI it had no oversight of a major scientific program that it had funded. Colwell and the Research Board were adamant that the Gulf of Mexico Research Initiative be motivated by scientific inquiry rather than private interests. The challenge was to make the mission clearly understood by the public and BP. BP's participation in or communication with the Research Board could jeopardize the scientific endeavor from the start and violate public trust. Hence, the Master Research Agreement and ethics policies kept GoMRI independent from its corporate sponsor.

Keeping GoMRI independent from BP was also a way to garner interest in its funding from the scientific community. When GoMRI initiated research funding, there was some concern on the part of Research Board members that the GoMRI funding would not be valued as highly as funding from federal agencies, such as the National Science Foundation or National Institutes for Health, because of a perceived relationship between GoMRI and BP. Hence, Colwell established, through committee, a rigorous ethics policy for GoMRI, discussed in detail in the following chapter.

While the ethics policy would resolve the issue of how scientists would see GoMRI funding, the question of how the public would view its activities remained. Beginning in 2014, through coordination with the Sea Grant program, which comprises university-based programs partially sponsored by NOAA, the Research Board assessed how the public perceived scientific research that was done in the Gulf. As described below, local communities in the Gulf of Mexico region sought scientific advocates, individuals, groups, or entities who could research problems and propose solutions to the oil spill that had devastated their local economy. In the first weeks and months after the Deepwater Horizon disaster, many community members expressed concern that politics would dominate the research agenda.[45]

One way to overcome public ambivalence about a BP-sponsored scientific research endeavor was through transparent management. While the high ethical standards that the Research Board put in place helped, there was a consensus among members that allowing the public to freely access GoMRI-funded data and research findings would also help foster greater community trust. GoMRI demonstrated its commitment to the public interest by mandating all researchers receiving funding publish their acquired data as soon as possible but no later than within one year of the closing of their grants. To meet this directive, GoMRI required researchers to submit their data to the Gulf of Mexico Research Initiative Information and Data Cooperative (GRIIDC). The GoMRI data management system provides public access to information gathered by researchers in their communities. Using GRIIDC, scientists collected, stored, and published data quickly, building on information gained from previous oil spills. The open research conducted with GoMRI funding alleviated some of the initial concerns of the media and the general public about the relationship between BP and GoMRI. Chapter three details the history of GRIIDC and GoMRI's data management policy.

The Shadows of Ixtox and *Exxon Valdez*

During the aftermath of the Deepwater Horizon platform explosion, the Coast Guard, NOAA, and other federal and state agencies enlisted scientists to measure the flow of oil from the broken wellhead. BP's initial estimate was "a bit on the optimistic side," according to William R. Freudenberg and Robert Gramling in *Blowout in the Gulf: The BP Oil Spill Disaster and the Future of Energy in America*.[46] Initially BP claimed that only about 1,000 barrels of oil a day were flowing into the Gulf, which would turn out to be only 2 percent of the actual volume leaking from the wellhead. Days later, independent scientists used satellite data to revise the estimate of the spill rate upward to about 5,000 barrels a day. In mid-May, National Public Radio released yet another estimate by independent scientists that placed the spill rate at about 50,000 barrels a day, an order of magnitude larger than what was first reported. The final estimate of the spill volume, reported as the well was finally being plugged, placed the amount of oil spewing from the well at 62,000 barrels a day initially, but declining over the course of four months to 53,000 barrels a day. This meant that for the duration of the spill the Macondo well—the official name for the well that was being bored from the Deepwater Horizon platform—had leaked about 4.9 million barrels of oil into the Gulf of Mexico, much of which was driven by winds and tides toward the beaches of the Gulf Coast states.[47]

In the days following the Deepwater Horizon explosion, the press, the public, and community members clamored for information about the Deepwater Horizon oil spill and attendant consequences. The public especially needed scientists to step in and determine exactly how much oil was escaping the damaged wellhead and where that oil was flowing. Any major oil spill in the Gulf of Mexico threatens sources of income and revenue, including fishing, crabbing, shrimping, and tourism. BP, on the other hand, had an interest in limiting information about the spill and underreporting its spread to avoid future liability. In the end, BP's public relations effort would alienate the general public, especially the Gulf community. GoMRI emerged in a contentious environment, where different stakeholders had contrasting ideas about who to trust and which data were reliable.

While the need to measure accurately the amount of oil escaping from the damaged wellhead was certainly drawing some scientists into the field, many oil spill researchers had other reasons for wanting to begin data collection and analysis as soon as possible. At the time, leading oil spill scientists compared Deepwater Horizon to the Ixtoc I oil spill, which occurred in the southern Gulf of Mexico from June 3, 1979, until March 23, 1980. Ixtoc was an exploratory well drilled by Petróleos Mexicanos (PEMEX), Mexico's state-owned and -operated oil company. Although geologically shallower than the Macondo well, reaching only about 11,000 feet under the Gulf—Macondo reached over 18,000 feet—and located in much shallower water, PEMEX's exploratory well posed many of the same problems as the Macondo well, including kicking and structural instability around the well bore.[48] As the PEMEX crew tried to cap the well, oil gushed from the wellhead, leading to nine months of spillage that released a total of 475,000 metric tons, or 3.4 million barrels, of oil

Figure 1.2. Aerial photograph of the Ixtoc I oil spill released by NOAA in September 1979. Collection of Doug Helton, NOAA/NOS/ORR.

into the Gulf.[49] That amount was slightly less than the total volume of oil released during attempts to plug the Macondo well, but the two were similar in their effects and consequences.

In the days and weeks following the Ixtoc I spill, slicks continued to flow into the northern Gulf of Mexico. By late July 1979—two months after the spill began—oil reached the shores of Texas.[50] In response, NOAA requested approval from the Mexican government to send a research vessel to investigate the oil spill site and affected areas along the Mexican Gulf coast, finishing its voyage in the Port of Galveston, Texas. The effort was part of a larger scientific survey of the spill and its effects that included other federal and state agencies, academic institutions, and private companies.[51] In his 2010 testimony to the National Commission on the BP Deepwater Horizon Oil Spill and Offshore Drilling, Dr. John Farrington, a participant on the Ixtoc research cruise and later a member of the GoMRI Research Board, contextualized the extent of US involvement in Ixtoc I research efforts.[52] The Ixtoc I spill occurred well within Mexican waters, and the government of Mexico decided not to grant permission for American vessels to simply enter their sovereign territory and conduct research on whatever topics they wished. Mexico had its own research agencies that could study the spill; the bulk of the American scientific work was done in US waters, predominantly along the Texas coast and far from the epicenter of the spill. Furthermore, the technology available at the time for monitoring the spread of oil was unsuitable for a spill the size and scale of Ixtoc I. As Farrington put it in his testimony, "the Mexican government did not grant permission for biological effects studies. A detailed physical oceanography study was not possible because of insufficient time to secure appropriate equipment for the cruise and also concerns about the irreversibility of oil damage to sensitive instruments or inability of the instruments to operate in the oil contaminated environment."[53] In the United States, the general public, federal and state governments, and the larger research community placed immense pressure on scientific agencies to investigate Ixtoc thoroughly, but various constraints and international borders limited the extent of US involvement.[54]

In his testimony, Farrington pointed out that the government of Mexico limited American participation in research efforts in the southern Gulf of Mexico. This was a plain statement of fact, based on the rights of a sovereign nation to control what parties cross its borders. Other scientists expressed more frustration with the scope of American research efforts when recounting Ixtoc I. Dr. Steve Murawski, a researcher specializing in population dynamics and marine ecosystems at the University of South Florida College of Marine Science and a recipient of GoMRI funding, claimed that, in the wake of Ixtoc I, the Mexican government, acting through PEMEX, "put a big foot down" on data and research, deemphasizing the need for transparency in the scientific research done in the Gulf. Only recently, Murawski claims, did the partnerships between PEMEX and foreign oil companies begin to hold the Mexican oil producer to higher standards of transparency.[55] In 2007, Jan Erik

Vinnem, an offshore risk management specialist at the University of Stavanger, Norway, wrote that the lessons learned from the Ixtoc disaster were 'unknown.'"[56] In 2010, comparing the Deepwater Horizon spill to Ixtoc I, *Reuters* reporter Robert Campbell maintained that "PEMEX never revealed the exact cause of the accident."[57]

Scientists have acknowledged that there remained unanswered questions following the Ixtoc I spill, but whether those gaps in knowledge were products of political decisions, the state of tools and techniques for data collection and monitoring, or the unprecedented size of the spill following the blowout remain matters for debate. In any case, the scientific community agreed that a long-term research and monitoring program like GoMRI, which addressed the role of oil in the Gulf of Mexico decades after a similar accident, could greatly improve the state of scientific knowledge. When the Deepwater Horizon explosion and spill occurred, Ixtoc I cast a long shadow over the scientific community. The lessons that scientists had learned about political expediency, the role of oil companies in limiting investigations and research, and the need for greater knowledge about the effects of large-scale oil spills on a range of ecological, environmental, and human consequences made strong cases for building an independent operation such as GoMRI.

Of course, Ixtoc and Deepwater Horizon are not the only two oil spills that linger in the minds of oil spill scientists. The Taylor Energy oil spill off the southeast coast of Louisiana, which was mitigated (but not stopped) in 2019 has leaked continuously into the Gulf for twenty years, as of 2024.[58] Other spills—both major and minor—have also affected important bodies of water. Teresa Sabol Spezio, a professor of environmental analysis, has traced the role of the 1969 Santa Barbara oil spill, caused by a major blowout on an offshore oil rig, in spurring federal lawmakers to finally pass long-delayed environmental legislation, transforming the regulatory landscape.[59] As she acknowledges, the Deepwater Horizon disaster gave coastal California residents a sense of *déjà vu*. In the opening symposium

Figure 1.3. Workers care for an oiled bald eagle and its chick (which is the blurry clump of feathers held by the two workers) in the aftermath of the Exxon Valdez spill. Images like this—often of waterfowl and other sea creatures completely covered in oil—were broadcast on nightly newscasts as responders rushed to contain the spill and steam clean oil from affected beaches. EXXON VALDEZ Oil Spill Trustee Council.

to the final Gulf of Mexico Oil Spill and Ecosystem Science conference in 2020, Farrington highlighted the 1957 Tampico Maru spill, which dumped dark diesel oil into the waters of the Pacific off the coast of California and the North Atlantic Torrey Canyon spill of 1967, which "remains a wakeup call" within the oil spill community.[60]

Perhaps no spill looms larger for the scientific community in the United States, however, than the *Exxon Valdez* disaster. On March 24, 1989, the *Exxon Valdez* tanker ran aground in Prince William Sound, Alaska, spilling 240,500 barrels of oil into the water. For months first responders, scientists, and federal agencies worked to burn, scrub, contain, and study the spilled oil, all while images of oil-coated fauna were broadcast to sympathetic viewers across the country.

By the time of Deepwater Horizon, however, scientists had more advanced and precise means of collecting data about oil spills. The disaster offered a chance for scientists to improve on the techniques of the past. Furthermore, through GoMRI the scientific community would have the funding necessary to study the short-, medium-, and long-term effects of the spill and a broad framework for research and improved data management policies. Clearly, many in the oil spill community were eager to put the Gulf of Mexico Research Initiative to work in the first weeks and months after the Deepwater Horizon spill. Previous encounters with major oil spills informed their sense of urgency.

Responding to the Deepwater Horizon Spill: Dispersants and the Black Box of Immediate Consequences

In the immediate aftermath of the disaster, the need for a quick response to protect the Gulf led to some actions that were not thoroughly vetted by scientists based in universities.[61] The most pressing issue was how best to apply chemical dispersants to the oil spill. Dispersants work by breaking oil into smaller droplets, which helps the oil dissipate into the water column rather than lingering on the surface. Crucially for flora and fauna near the surface of bodies of water, dispersants reduce the slicks that would otherwise cover beaches. Dispersants do not remove oil completely. That is only achieved as the chemical bonds holding oil droplets together are exposed to sunlight, which degrades the small oil droplets over a very long period.[62] Dispersants had long been used during large-scale oil spills, but almost solely on the surface of the water.

From April 27 until May 3, first responders sprayed about 140,000 gallons of dispersants on the surface of the Gulf of Mexico. That figure was about two-and-a-half times the total amount of chemical dispersant used during the Exxon Valdez oil spill. During the following week, responders applied more than 165,000 gallons of dispersants.[63] In the weeks that followed, responders would distribute approximately 2.1 million gallons of dispersants in total.[64] Scientists had been debating the appropriateness and safety of dispersants since their introduction at the BP oil spill site. Some workers who

Figure 1.4. The US Air Force assists in Deepwater Horizon spill response activities. In this image, a C-130 Hercules drops oil dispersant on the surface of the Gulf to break up the oil slick. US Air Force Photo / Alamy Stock Photo.

handled and transported the dispersants reported that they had nausea and headaches, alarming Gulf Coast residents about the threat posed by the widespread distribution of chemical dispersants.[65] Some researchers also worried that dispersants could potentially harm the larger Gulf ecosystem, since tremendous amounts of oil and dispersant mixture were left floating in the water column. Scientists simply did not know how the unprecedented scale of dispersant application could affect the Gulf ecosystem.

By late April 2010, BP began requesting approval from the Coast Guard's Unified Command to inject chemical dispersants directly at the wellhead. This was an unprecedented request, and the dangers of using those compounds in such a way were unknown. There was simply too little published scientific research at the time on the effects of chemical dispersants used underwater. Some private companies had conducted limited research, but the background information was lacking. Ultimately, however, after establishing testing protocols developed jointly by NOAA and BP scientists, EPA

Administrator Lisa Jackson approved the use of subsea chemical dispersants at the damaged Macondo well. On May 10, 2010, Jackson allowed BP to apply up to 15,000 gallons of dispersant a day at the wellhead, given that appropriate steps would be taken to ensure that the toxicity of the chemicals would be limited. In the interest of mitigating the flow of oil to the surface, BP and the EPA took a risk, uncertain of the payoff. GoMRI would later dedicate part of its scientific mission to studying the effects of oil-dispersant mixtures on the Gulf environment and ecosystem.

Securing Public Trust

The Deepwater Horizon oil spill had spurred a deep mistrust on the part of the Gulf community toward anyone associated with BP. Adding to that mistrust was the revelation in May 2010 that BP either underestimated or underreported the amount of oil that was spewing from the broken wellhead daily by almost 55,000 gallons. Furthermore, the actions of BP's chief executive officer, Tony Hayward, had done little to endear the public to the company in the months following the spill, as he offhandedly claimed that "the Gulf of Mexico is a very big ocean, and the volume of oil we are putting into it is tiny in relation to the total volume of water."[66] That description of the spill did not fit with the images and videos that the public saw daily from the broadcasts of the broken wellhead and clean-up efforts on the surface. On his apology tour through local Gulf communities, Hayward once commented, "I just want my life back."[67] Hayward would have his wish one month later, when he tendered his resignation to the BP Board in July 2010.[68]

Despite Colwell's acknowledged expertise in biology, microbiology, and the physical sciences, some members of the press expressed doubts about the degree of independence the research initiative would have from BP. Pete Spotts, a reporter for *The Christian Science Monitor*, commented that the Natural Resources Defense Council, in particular, was apprehensive about the way that the initiative was being developed. Rather than going through a panel established by someone chosen by BP, executive director Peter Lehner argued, "the advisory panel should be named by an organization such as the National Academy of Sciences."[69] Spotts also noted, however, that there were "what appear to be layers insulating individual researchers from taking a direct hand-off from BP's pay master."[70]

Since the first "quick-start" grants to research institutions had gone out in June 2010, reporters also called into question the ways in which funds were being disbursed, searching for any possible connection between BP and scientific research conducted through their grant money. In response, Carron stepped in to calm the media and general public, stating that "there's no pre-approval on the research we're going to fund. And there's no pre-approval on any of the publications."[71] By preapproval Carron was referring to oversight or control on BP's part over the initiative's choice of funding recipients. The decision of who to fund would be made entirely by the Research Board. Furthermore, BP required that all research results had to be published in independent, peer-reviewed journals. The company did wish to be notified when any publications were about to be released, but BP exerted no editorial control.[72] As detailed in the following chapter, the Research Board would rely on robust ethics policies to assure the public that GoMRI science was accurate and trustworthy.

Conclusion

The Gulf of Mexico Research Initiative provided a novel approach to the funding, management, and public accessibility of scientific research and knowledge. First, the organization was established in direct response to a major oil spill. With the legacy information gaps in the wake of Ixtoc I and *Exxon Valdez* still in the minds of prominent oceanographers, marine scientists, engineers, and other experts, there was much pressure within the scientific community to investigate fully the short-, medium-, and long-term consequences of the oil spill and the decisions made in response to the disaster.

Second, although created using industry funds, GoMRI was able to establish itself as an independent scientific organization, conducting research but also disseminating that research to the general public and using the scientific knowledge to understand and assist the Gulf community as it recovered from the Deepwater Horizon disaster. The Gulf of Mexico Research Initiative also encouraged the general public and researchers unaffiliated with the initiative to utilize the research, conclusions, and data generated through GoMRI funds for purposes in the public interest.

Third, GoMRI's public-facing nature was a feature built in to the organization from the outset. Dr. Rita Colwell played a central role in developing a research initiative that engaged with the public at large from the time that she was approached by BP. Through her efforts and those of the Research Board, GoMRI became a successful research funding organization, advancing independent, peer-reviewed, and effective scientific research in the Gulf of Mexico. These factors—GoMRI's public orientation, its efforts to improve on the state of knowledge about oil spills in the long aftermath of Ixtoc I and *Exxon Valdez*, the ways that GoMRI successfully distanced itself from BP, and the organization's work on behalf of the public and other stakeholders—made the initiative significant in the history of science and scientific funding.

CHAPTER TWO

DISTRIBUTED MANAGEMENT
Administering Scientific Research
During an Unfolding Crisis

The Gulf of Mexico Research Initiative (GOMRI) benefited from the exceptional leadership experience of the Research Board chair, Professor Rita Colwell, and Vice-Chair Professor Margaret Leinen. They were able to assemble a Research Board for overseeing the program despite the initial political machinations resulting from the pledge of $500 million over ten years made available by BP as explained in chapter one. Strict adherence to the Master Research Agreement enabling GOMRI and the Research Board By-Laws prevented any interference with Research Board decisions by BP, U.S. federal, state and local political entities. There was strict adherence to Conflict of Interest Rules by members of the Research Board.

The wide-ranging scientific research, education and leadership experience of the Research Board members facilitated GOMRI partnership building with existing organizations that enabled the administrative support needed while minimizing administrative costs. The

Research Board utilized an excellent Leadership Team of Chief Scientific Officer Chuck Wilson, Program Manager Kevin Shaw, and Program Director Mike Carron to lead and manage the day-to-day operations via a "distributed management process." The grant process from evaluation reviews, through awards, actual research and outreach, peer reviewed publications and grants closures very much resembled grants and research funded by the National Science Foundation and other similar U. S. federal agencies.

All of the above enabled research published in peer-reviewed scientific publications that advanced significantly our understanding of the Gulf of Mexico ecosystems and the fates and effects of petroleum in the marine environment. Involvement of early career scientists—undergraduate and graduate students, and postdocs—enabled a cadre of early career scientists with experience at the interface of fundamental research and research with immediate societal needs. Public education, including K–12 grades, and outreach were much acclaimed.

Participation as a member of the Research Board was one of the highlights of my fifty plus year career in ocean sciences.

—DR. JOHN FARRINGTON, Woods Hole Oceanographic Institution[1]

This chapter provides a history of GoMRI's management structure—what some on the management team like to refer to as a distributed management model—and descriptions of its component parts. It highlights the ways that the initiative built a flexible and responsive organization by empowering different individuals and groups to take on important tasks using the new communication technologies available in the second decade of the twenty-first century. GoMRI was able to respond effectively and nimbly to the managerial and administrative challenges of building a research organization in the midst of an unfolding disaster. Decentralizing many of its day-to-day operations was crucial to its rapid development. Through the efforts of many different individuals and institutions, GoMRI presented a viable model for how to marshal the scientific community in response to a major disaster.

The Research Board relied on each member and each consortium leader to pursue and achieve its overall objectives because it was not possible for every member of the board to be in all of the locations in which GoMRI operated. GoMRI's research model, like that of the National Science Foundation, allowed its consortia, small teams, and independent researchers to manage their projects independently. The Research Board used quarterly and annual reporting requirements, regular calls with team leaders, and consortia site visits to evaluate progress. Furthermore, GoMRI enlisted partner organizations and units—the American Institute of Biological Sciences, the Northern Gulf Institute, the Consortium for Ocean Leadership, and Sea Grant—to meet its goals as stated in the Master Research Agreement. The Research Board, GoMRI Management Team, consortia and small team leaders, and partner organizations worked collectively; each had different management responsibilities, yet they combined their efforts and aimed toward making the resultant research successful.

Rather than assembling organization managers and team leaders into a physical office, distributed management worked by having most interactions between individuals online or on the telephone. Fortunately, the Research Board organized GoMRI at a time when tools for virtual communication were becoming widely available and better suited for virtual office work.[2] By taking advantage of virtual meetings and office spaces, GoMRI was able to weather unforeseen difficulties. For example, the Gulf region is susceptible to violent storms almost all year, although it experiences a higher rate of disasters during hurricane season in the summer and fall, as the warm,

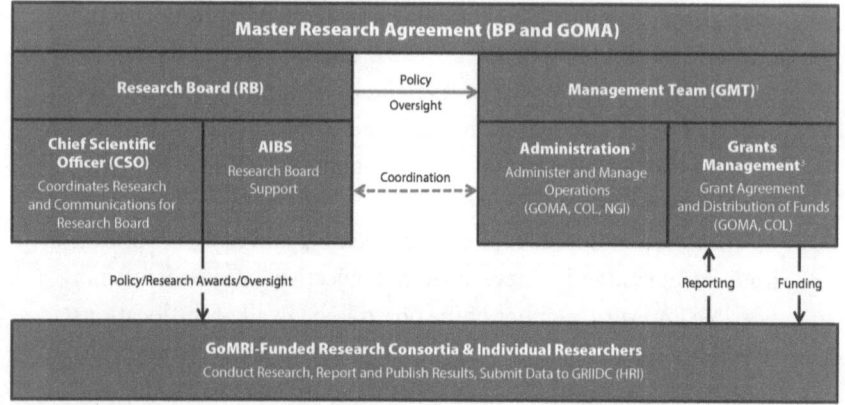

Figure 2.1. The chart above depicts the organizational structure of the Gulf of Mexico Research Initiative. From Leigh Zimmermann et al., "From Disaster to Understanding: Formation and Accomplishments of the Gulf of Mexico Research Initiative," *Oceanography* 34, no. 1 (March 2021), 19. Licensed under CC BY 4.0, https://creativecommons.org/licenses/by/4.0.

moist air from the Gulf of Mexico fuels immense storms. Those storms sometimes have the potential to devastate office buildings that house research centers located along the Gulf Coast and further inland. Hurricane Harvey, for example, threatened the Harte Research Institute, described in the following chapter, as the storm bore down on the Texas coast in August 2017. Harvey created some concern among members of the Research Board about the physical security of GoMRI's data storage systems located in Corpus Christi at the HRI.[3] The GRIIDC staff were able to quickly put together a plan to protect the data, which thereby maintained GoMRI's larger commitment to free and open data.[4] All of the planning and preparation for the storm took place in virtual space among members of the GoMRI Research Board and management team.

GoMRI's management system was not purposely implemented as a disaster mitigation plan, although it acted as such at certain moments. Rather, distributed management emerged from the fact that, during the months of the spill and immediately after, the GoMRI Research Board needed to build a substantial scientific funding program very quickly. Time was a luxury that the board could ill afford as scientists hurried to enter the field and the public clamored for information and knowledge about the state of the Gulf. The Research Board had to get projects off the ground as soon as possible during 2010 and 2011.[5] It needed to be flexible in its response and did not have time nor did it wish to build a large in-house administrative and support staff. Furthermore, the sunset clause in the Master Research Agreement meant that, with exceptions, the initiative would cease to function after 2020. While GoMRI might have had an end date, the organizations with whom GoMRI partnered would continue research on the role of oil in the Gulf of Mexico long after 2020.

Function of the Research Board in a Distributed Management System

Standing at the top of the GoMRI administrative structure was Dr. Rita Colwell, chair of the Research Board for the entirety of GoMRI's existence from 2010 until 2020. Her role as chair began when she accepted Dr. Ellen Williams's phone call on May 26, 2010. For a decade, Colwell steered GoMRI and ensured the smooth disbursement of $500 million dollars for responsible, transparent, and rigorously reviewed scientific research.

The Master Research Agreement gave specific powers to the chair of the Research Board. First, the chair presided over all of the meetings of the Research Board, and was responsible for preparing agendas, minutes, reports, and meeting summaries. Second, to ensure continuity of the board's operations, the chair also had to appoint a vice chair from among the Research Board members who could act in the chair's stead when necessary. Colwell selected Dr. Margaret Leinen, an accomplished oceanographic scientist, to fill the role. The two had a long professional history, including several years serving together on the NSB and at NSF. Colwell was also impressed by Leinen's service as the vice provost for Marine and Environmental Programs and dean of the Graduate School of Oceanography at the University of Rhode Island.[6] Third, the chair acted as the official source of communication between the Research Board and any third party, including BP, the press, and the public. CSO Chuck Wilson assisted Colwell with that responsibility. Finally, the chair was responsible for exercising powers permitted by the by-laws or resolutions of the Research Board.[7]

The GoMRI by-laws, which were additional policies that clarified, strengthened, and added to the governance structure dictated by the Master Research Agreement, did place some checks on the chair's authority. To ensure rapid build-out of GoMRI, Colwell was retroactively confirmed as the inaugural chair in the by-laws. This meant that the Research Board did not have to be concerned about appointing Colwell chair in 2011; by nature of her role in the first months of GoMRI she was officially installed in the position. The by-laws added both a term for the chair of the Research Board and a process for selecting a new chair. The chair's term was limited to three years, with consecutive terms permitted.[8] Although the by-laws stated that the chair was subject to a three-year term, the Research Board retained the utmost confidence in Colwell and no candidates came forward to run for the position at any time during GoMRI's history.[9] The Research Board had simply concluded going through a nomination process every three years would be a waste of valuable time since the lifetime of the initiative was only ten years.

Borrowing some of the best practices of the National Science Foundation, Colwell and the Research Board crafted by-laws and a code of conduct to ensure the initiative would be a transparent, reputable, ethical, and trustworthy scientific funding organization. Members of the GoMRI Research Board were personally ineligible for GoMRI funding. Furthermore, no board member could participate in the review or approval of a proposal that might create a conflict of interest. Research Board members were barred from participating in discussions about grants that could have been awarded to their home institutions or partner organizations to their home institutions. This helped eliminate potential accusations or even an appearance of favoritism in the distribution of research funds.[10] As noted above, scientists may strive for total impartiality, but in reality that is nearly impossible. What scientists can do, instead, is adhere to codes of conduct and ethics that limit to the greatest extent possible any bias, actual or potential, in their decision-making.

Members of the Research Board were not employees of the Gulf of Mexico Research Initiative. They did receive an annual honorarium as renumeration for their labor, but it was not enough to serve as their primary income. While serving on the board they continued their other professional responsibilities. Most board members were full-time teachers, researchers, or administrators employed by universities and research institutions.[11] The Master Research Agreement clearly stipulated that Research Board appointees had to be independent scientists or administrators whose professional work would not compromise their ability to make impartial decisions on behalf of GoMRI.[12] Dr. William T. Hogarth's adherence to the conflict-of-interest policy came up the most among the twenty Research Board members. Hogarth served as the director of the Florida Institute of Oceanography (FIO) during his time with GoMRI. The FIO is an entity of the state university system of Florida and works with twenty organizations and agencies across the state. Either the FIO or its partner organizations sought funding from GoMRI on every RFP solicitation round. Hogarth was required to leave the room so many times during proposal reviews that other board members humorously quipped "is Bill in the room?" at the start of every proposal review meeting.[13] While amusing, the question about Hogarth's presence during RFP reviews underscores how seriously the Research Board took its code of ethics.

To address the significant aspects of its research mission, the board formed committees, each of which was tasked with handling a particular issue.[14] For example, the data management committee ensured GoMRI grantees abided by the mandates of the Master Research Agreement, notably to submit data in a timely manner. It also proposed and implemented subsequent clarifications to that policy.[15] Generally, new committees were formed when one or two members of the Research Board identified the need for special attention to a particular issue. For example, the book committee consisted of a few members of the board who identified the need for a book documenting the history and achievements of the Gulf of Mexico Research Initiative. Rather than overseeing each committee directly, the chair appointed a Research Board member, who may have suggested a committee be formed, as that committee's leader. By restricting committee membership to three to six people, with rare exceptions, the Research Board ensured that only interested and knowledgeable individuals would participate in any given committee. Furthermore, by dividing work on various components of GoMRI into committees, meetings consisting of the entire Research Board were spent ensuring GoMRI progressed toward its goals in a timely fashion. After internal deliberations, committees presented recommendations pertaining to policy or practice to the full board for discussion and formal vote. Delegation was central to how the Research Board handled the larger issues, and it allowed board members to focus many different tasks simultaneously.

While the Research Board constituted the primary administrative and deliberative organ of the GoMRI leadership, the management team supported the initiative's scientific mission in significant ways. Three individuals within the management team—Chief Scientific Officer Chuck Wilson, Program Manager Kevin Shaw, and Program Director Mike Carron—comprised the leadership team. Tasked with ensuring timely achievement of project goals, the leadership team engaged in the direct, day-to-day management of the Gulf of Mexico

Research Initiative. Beginning in 2011, Chuck Wilson, Kevin Shaw, and Mike Carron collaborated closely to discuss the general direction of the organization operations. Their expertise allowed them to provide comprehensive overviews of the organization to the Research Board at any given time. The leadership team provided information about the functioning of any part of the initiative, from the level of the individual researcher, each consortium, and every partner organization, at any point in time, thereby assisting the Research Board in deciding where its attention needed to be directed. Their roles, therefore, are described below in additional detail.

The Leadership Team: The First Administrative Layer of GoMRI

The chief scientific officer served on both the management team and *ex officio* on the Gulf of Mexico Research Initiative Board. In this role, Wilson was assigned three primary tasks: serve as communicator between the Research Board and BP; keep the Research Board informed of research activities; and oversee execution of Research Board decisions. The CSO was a full-time employee of GoMRI with responsibility to maintain contact between the different components of GoMRI's scientific program and to act as liaison between staff, researchers, and the Research Board. Wilson's work was facilitated by a stipulation in the GoMRI by-laws that the CSO served as an *ex officio* member of every Research Board committee.[16] Wilson's broad mandate meant that he maintained an understanding of the initiative from the high-level operations of the board to the intricacies of how each sponsored project was working to complete its goals. He was in daily contact, with some exceptions, with the GoMRI Board chair and in close contact with researchers, management staff, and outside contractors brought into GoMRI for various assignments.

The chief scientific officer also played an important role in maintaining social and institutional distance between BP and GoMRI. Initially, Colwell and the Research Board were in a tenuous position between an aggravated and concerned local community seeking answers about the health and quality of the Gulf of Mexico and the private company that was widely blamed for the spill. In order to both streamline communication with BP—which was interested in the operations of GoMRI in its formative years—and to keep the company at an arm's length, Wilson took on the job of communicating with BP and the Research Board. According to Wilson, his job as a liaison consisted solely of informing a counterpart at BP about research that GoMRI was funding and other matters as directed by the Master Research Agreement.[17]

After appointment as chief scientific officer in 2012, Wilson became a vital part of the GoMRI management structure.[18] He and Colwell were public faces of the organization. At the final Gulf of Mexico Oil Spill and Ecosystem Science (GoMOSES) conference in February 2020, Wilson was awarded the Wes Tunnell Lifetime Recognition for Gulf Science and Conservation award. The ceremony commemorated Wilson's long career spent researching in his field, notably in the Gulf of Mexico, and administering programs to better understand the Gulf, its ecosystem, and the human relationship with the sea. Dr. Larry McKinney, senior executive director of the Harte Research Institute and chair of the GoMOSES Executive Committee in 2017, 2018, and 2019, highlighted Wilson's work within GoMRI during his presentation of the award.[19]

In addition to Wilson, Kevin Shaw, the GoMRI program manager, often served as one of the first points of contact for new consortia, individual researchers, contractees, and partner organizations to GoMRI. Shaw also had deep personal connections to some of the individuals involved in the formation of the Gulf of Mexico Research Initiative. He had attended graduate school with McKinney, who kept him in the loop about the BP financial commitment and during early drafting of the Master Research Agreement in 2010.[20] Shaw was interviewed by Mike Carron, program director, and Research Board member Dr. David Shaw in late 2010, and was brought on in 2011 to develop and guide the GoMRI management team.

While working to define his own role within the management structure of the organization, Shaw was responsible for finishing the first draft of the Master Research Agreement between GOMA and BP. Together with Carron and several other partners, Shaw helped develop and oversee the distributed management system. With only the Research Board and a handful of staff in place in early 2011, Shaw and Carron looked to partner organizations to help get GoMRI up and running. By October 2011, Shaw and Carron had solidified a three-party agreement among GOMA, BP, and Ocean Leadership that clarified how funds to research grantees would be distributed throughout each year.[21] Shaw understood that building a large research funding apparatus would require assistance from established institutions. Bringing in other parties would also build personal and professional connections that would outlast the initiative, as discussed in the following chapters.

As with Wilson and Shaw, Carron had several different responsibilities. Although his title was director of the Gulf of Mexico Research Initiative, he played a key role in drafting the MRA, creating GoMRI financial policies, and administering the initiative as a whole. Prior to joining GoMRI, Carron had spent his career conducting scientific research and administering research programs, beginning with the United States Naval Oceanographic Office (NAVOCEANO), where he worked on mathematical models of ocean climatologies and geophysical and oceanographic data analysis. By the 1990s, Carron had been appointed chief scientist at NAVOCEANO, responsible for overseeing and administering global oceanographic surveys. In the 2000s, Carron moved to NATO, where he focused his research on the effects of sonar and artificial sounds on marine mammals. After five years with NATO, Carron was appointed chief scientist and, ultimately, director of the Northern Gulf Institute, a consortium of several academic and research organizations focused on oceanographic and geophysical studies of the waters of the Gulf of Mexico that were under United States jurisdiction.[22] Carron's career prior to his work with GoMRI represented a blend of applied scientific research and project management, vital skills that fit appropriately within the GoMRI approach to funding scientific research. Carron's contributions to the GoMRI operations, and the fact that his position required knowledge about top-level decision-making, resulted in his serving as *ex officio* Research Board member.[23]

Carron developed many of the policies pertaining to how the BP funds would flow and focused most of his effort overseeing financial planning and management. During the first months of the initiative, he helped draft the Master Research Agreement. Ultimately, Carron's budgeting expertise and the reliance of GoMRI on relatively inexpensive video and telephone technologies meant that the cost to GoMRI on overhead was significantly less compared to other scientific funding agencies. GoMRI expenditure on overhead constituted only about 10 percent of the total budget, which maximized funds for scientific research.[24]

The leadership team was not a de facto group within the GoMRI management structure. There is no such label shown in figure 2.1. Its existence was organic, with Wilson, Shaw, and Carron realizing their responsibilities to the initiative were similar in scope and sometimes overlapped. All three had to ensure that every component of GoMRI—from individual scientist to partner organization involved in the initiative—advanced the organization's scientific goals. The history of the leadership team reveals that the GoMRI distributed management was flexible, bringing talented researchers and administrators together and allowing them to build the initiative program. The leadership team's modus operandi and ways that its members carried out their responsibilities exemplifies the flexibility of GoMRI in pursuit of its scientific mission.

GoMRI Model for Scientific Research: Distributed Management with Strong Coordination

Responsibility for managing scientific research fell to the scientific investigators. This distanced the board from the researchers and allayed concerns that corporate or political interests were interfering with the research. The Research Board approved the best proposals for funding and left the work of managing research projects to each consortium and small-team lead investigators. Consortia consisted of a principal investigator (PI) and co-PIs located at three or more institutions. The small-team investigators consisted of either a single PI or one PI with up to three co-PIs. According to board member Debra Benoit, this was similar to an NSF cooperative agreement, in which GoMRI offered a span of support, rather than a span of control. This allowed the consortia to manage their own research program.

With the one exception of RFP-III, GoMRI issued periodic RFPs for research consortia or small teams.[25] RFP-I and RFP-IV were issued for consortia proposals in 2011 and 2013. RFP-II and RFP-V, released in 2011 and 2014, solicited principal investigators who planned to work with up to three co-principal investigators in small teams. RFP-VI, released in 2016, combined the consortia and small team RFPs in a single solicitation. A timeline showing the RFP release dates, as well as the amount of money awarded in each round, can be found in Appendix E.

GoMRI chose to issue funding to consortia and small teams for several reasons. By funding consortia, GoMRI built connections among institutions throughout the United States and internationally, but predominantly in the Gulf of Mexico region. This would address GoMRI scientific themes, contribute to interdisciplinary research, and augment the Gulf region capacity for scientific research. Capacity building, discussed in chapter four, was of interest to both BP and GoMRI, hence consortia were excellent vehicles for building scientific capacity in the Gulf region. RFP-II and RFP-V were issued to solicit research ideas from individual and small-team investigators. Furthermore, based on their experiences reviewing proposals for the National Science Foundation and other agencies, the Research Board members understood some researchers perform better on their own than with large interdisciplinary groups, hence the request for proposals from individual investigators and small teams.

The Research Board led the effort to draft RFPs. As with its other responsibilities, the Research Board relied on committees to draft and revise the initial language of each RFP before submitting it to the full Board for final approval. During the drafting process, the board worked closely with Ocean Leadership to ensure that the final language was in accordance with the best practices and standards of oceanographic funding solicitations and the MRA. Ocean Leadership, primarily represented by Leigh Zimmermann, was also recruited to assemble the final RFP packets and advertise them to appropriate institutions.

Once the Research Board finalized its funding commitment for the consortia and small teams, it established communication via annual and quarterly reporting and site visits. Management of scientific research was the responsibility of investigators, which was the standard practice of research funding organizations such as the NSF, NAS, and others. If investigators encountered problems executing projects outlined in their RFP submission, the chief scientific officer reviewed and presented the issue to the Research Board, and together they developed a strategy to move forward.

After the Research Board approved selection of an RFP, a contract officer was assigned who was responsible for administering each research project by overseeing contract negotiation, signing, ensuring the research team adhered to contract stipulations and monitoring timely submission of annual and quarterly finance and activity reporting. The contract officer also closed the project once it was completed. This had some advantages. Having one individual oversee many different contracts was cost effective and maximized the portion of funding directed to scientific research. Another advantage was that researchers had a single dedicated point person for questions about their contract, reporting, or data submission. To ensure contract officers were not overburdened, standardized forms were employed and information on each grant was made available in Microsoft SharePoint, where it could be easily accessed by both the management team and the Research Board.

The major objective for GoMRI was to produce the highest quality scientific research on effects of oil spills in the Gulf of Mexico region, but this also laid groundwork for future scientific undertakings. Although GoMRI was scheduled to be completed in 2020, the Research Board anticipated the initiative would enhance the research capacity of the Gulf region. By encouraging consortia to address one or more of the research themes and by bringing PIs and researchers together from around the world who may not have otherwise had the opportunity to collaborate, GoMRI forged interdisciplinary connections within the community of Gulf oil spill researchers, supporting professors, graduate students, and undergraduate students participating in the consortia. The effects of GoMRI's funding were not limited to the Gulf region alone; the consortium model encouraged partnerships with research institutions and universities across the United States and around the world. Although the vast majority of GoMRI funding was spent in the Gulf region, there was nothing preventing regional institutions from linking together with other campuses and research centers. As Dr. John Farrington has noted, this was especially the case for research focused on the fates and effects of oil spills, but also marine environmental quality more generally.[26]

GoMRI RFP Solicitation, Review, and Funding

The Research Board established committees for each of the six RFP cycles. Each committee worked with the editorial assistance of Ocean Leadership to develop each RFP and a framework for its solicitation and review. The entire board then convened to approve each RFP. The board reflected on what it had learned from past RFP reviews, and took care to modify subsequent RFPs as necessary, resulting in slightly different wording for each RFP cycle.

Each time the Research Board prepared to release a new RFP, it first put out a formal press release. To spread word about the

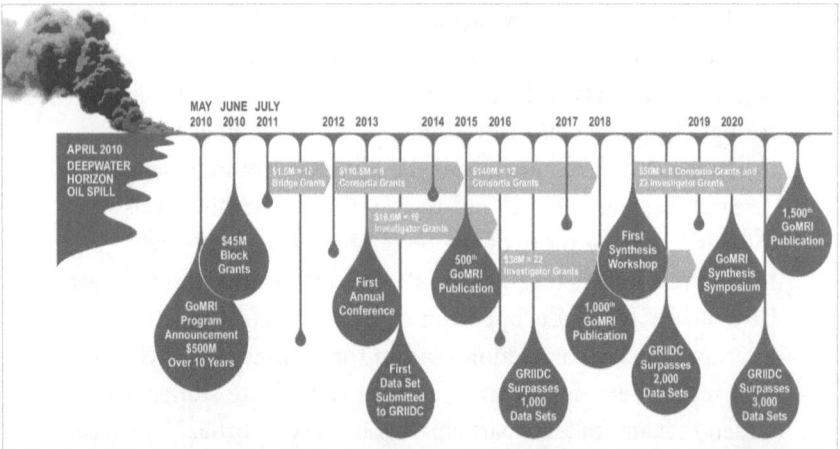

Figure 2.2. This timeline was first published in Leigh A. Zimmerman et al., "From Disaster to Understanding: Formation and Accomplishments of the Gulf of Mexico Research Initiative," *Oceanography* 34, no. 1 (March 2021): 16–29; https://doi.org/10.5670/oceanog.2021.114. It shows when different rounds of funding began and the years they were meant to cover research, along with other major milestones in GoMRI history. Licensed under CC BY 4.0, https://creativecommons.org/licenses/by/4.0/.

RFPs, the press release was posted on the GoMRI webpage and then emailed to potentially interested parties via Ocean Leadership, the Gulf of Mexico Sea Grant programs, and other organizations with relevant researcher databases. Each release pointed interested parties to the GoMRI RFP web portal that contained the full RFP and other necessary forms and details.

Once the RFP submission deadline passed, the board convened to review the documentation. Its review process ensured the best proposals—the most viable submissions and those capable of producing trustworthy, useful, and significant scientific knowledge—would be funded. The first step was an administrative review by Ocean Leadership, which examined submissions for adherence to the basic, RFP-specified requirements for a complete proposal. Proposals that were missing required elements were rejected in this preliminary round. The team at Ocean Leadership also ensured that proposals adhered to the maximum page count and would eliminate pages that exceeded the page count limit, sometimes resulting in unfinished sentences or paragraphs.[27] Fairness in the RFP process required that GoMRI hold all applicants to the same standards. RFP-VI, the final RFP, serves as a good example of what administrative compliance entailed. In addition to adhering to the definition of a consortium or small team,

> [t]he Lead Research Institution must be a US academic or nonprofit research institution. GoMRI research shall be led by PIs primarily in the Gulf Coast States; however, US investigators from institutions outside the Gulf Coast States are not prohibited from applying as PI or co-PI. Projects may include internationally-based co-PIs.
>
> The collaborating members or institutions may be drawn as needed from US and foreign universities and government laboratories, Federally Funded Research and Development Centers, and other US and non-US institutions with unique, world-class capabilities. An investigator may have an existing relationship or may find it advantageous to enter into a new relationship with a private or for-profit entity. Individuals at private or for-profit entities can contribute in the form of products, services, and expertise that will be crucial to the delivery of the research objectives. As with all participants in the GoMRI, any activities by government, private, or for-profit entities will be subject to the terms of the GoMRI MRA.
>
> An individual can be the PI or Consortium Director of only one submitted proposal to this RFP and only involved in up to three submitted proposals in any capacity (e.g., PI, Consortium Director, co-PI, collaborator, etc.). If an individual appears on multiple proposals, this should be clearly noted in the List of Participants and in Current and Pending Support; in each proposal, a clear description should be included to explain how the proposed work is complementary, not duplicative, of other proposed efforts and how the participant will budget his or her time. Should an individual be the PI or Consortium Director

on more than one proposal, or should an individual appear on four or more proposals, all proposals involving the individual will be disqualified. No individual should therefore be included without their explicit consent. It is the responsibility of the submitters to confirm that each member of the entire team is within the eligibility guidelines.

Each proposal must be a stand-alone document without possibility of linked proposals.[28]

If a proposal did not adhere to the required stipulations, it was rejected by the board. The administrative review ensured only those projects that met the standards outlined by the Master Research Agreement and the stipulations of the RFP solicitation were considered for possible funding by the Research Board. Once it had finished its initial administrative review, Ocean Leadership handed the proposals over to the Research Board and to external reviewers.

Proposals that cleared the administrative review were subjected to rigorous peer review, in keeping with the model provided by the NSF. Scientists with appropriate credentials and without any conflict of interest relevant to the RFPs served as external reviewers. Ocean Leadership used their reviews to rank projects according to weighted criteria. For RFP-VI, the criteria included: scientific value and merit (80 percent for individual investigators; 70 percent for the consortia); qualifications (20 percent for individual investigators; 15 percent for consortia); and management (15 percent for consortia only). Ocean Leadership submitted the results—compiled on a standardized scoring sheet for each project—to the Research Board to inform its final decisions. The Research Board operated with a fixed amount of money on an annual basis, being allocated a total expenditure of $50 million for each year from BP. Administrative costs for GoMRI, per year, were approximately 10 percent of the total amount awarded by BP. After allocating money for overhead and data management, the Research Board determined how much of the $50 million remained for scientific research. If not all of the total amount was spent in a given year, it could be rolled over to the next year. Consortia could also roll over unspent money into the following year since most RFPs covered two or three years of operations.

Unless they had a conflict-of-interest with an organization that submitted a particular RFP (such as being employed by the same institution as the applicant), Research Board members were given copies of the RFPs to review, along with external evaluations to read and critique. Research Board members were not expected to read every proposal, but special care was taken to ensure proposals for specific scientific themes were assigned to members with relevant expertise. Collectively, Research Board members possessed a range of qualifications and expertise across several fields and disciplines, ensuring a sufficient set of reviews for each proposal. Board members were given access to a list of all proposals that were not eliminated due to the administrative review process; nonconflicted board members were allowed to request any proposal, no matter its score from the external reviewers, to be considered by the full board, ensuring that no proposal would be inadvertently overlooked. As it finalized its funding decisions, the Research Board examined the mix of proposals to ensure that each cycle comprised a "balanced bouquet," the board metaphor for funding proposals across all scientific themes.[29]

Evaluation criteria for RFPs were like those of other funding agencies and organizations, like the NSF, EPA, NIH, and others. The RFP-VI score, for example, predominantly comprised scientific value and merit. A board member noted "scientific value and merit were indicated by the quality and innovative nature of the research as it applied to one or more of the GoMRI themes. If those criteria were met, then the issue of the scope, quality, and potential of the proposed public education and outreach activities became important in evaluating the proposal."[30] The Research Board encouraged applicants to detail the scope and significance of work and describe how their research would add value to one or more of the GoMRI research themes. If a consortium requested funding renewal, the RFP asked how continuing the funding would build on the scientific legacy of its previous effort.[31] PIs, co-PIs, and consortia directors were required to include in their proposal a history of professional activities, publications, demonstration of their knowledge of the Gulf of Mexico, and capacity to carry out the proposed project.[32]

As is the practice at NSF, the Research Board required each consortium to function independently. In turn, GoMRI fostered connections between and among consortia through conferences, workshops, and regularly scheduled webinars, in keeping with the Master Research Agreement call for a broad, interdisciplinary approach to scientific funding. According to the GoMRI distributed management model, the initiative provided connections among the different consortia, each of which managed its own scientific research.

Partners to GoMRI Research Efforts

Organized during an unfolding disaster, GoMRI had to act quickly to mobilize scientists and prepare them to do fieldwork. To accomplish these tasks, it partnered with organizations with a history of supporting oceanographic research in the United States. The fact that GoMRI brought new partners into its operations, as needed, reflected the evolution of Research Board priorities. The partner organizations became crucial components of the GoMRI management structure. The result at the end of the ten years is a legacy that will outlast GoMRI.

Gulf of Mexico Alliance

Before funds could start flowing to scientists and their research institutions, the primary challenge was to determine how money would be disbursed from BP to GoMRI. With oil continuing to gush from the seafloor, Colwell and the early Research Board had little time to form an entirely new fiduciary agent to handle the $500 million commitment. The early leadership of GoMRI recognized the scientific community needed to respond urgently to the Deepwater Horizon disaster, namely by collecting samples for early phase research.

GoMA's role in the formation of GoMRI is discussed in detail in the previous chapter. After the early negotiations among the federal government, Gulf state governors, BP, and the Research Board, GOMA continued to play a key role as a financial agent for GoMRI. The GOMA partnership with GoMRI created an additional layer of oversight, audit, and review for financial disbursement and accounting. As discussed in the previous chapter, the public—especially the Gulf community—was skeptical of organizations accepting money from BP. In the summer and autumn of 2010, Colwell and the Research Board took necessary steps to ensure BP would not influence the selection of scientists receiving funding. GOMA, experienced in funding, coordinating, and supporting multistate programs to study the Gulf of Mexico, insulated GoMRI from any undue influence from BP in the selection of scientists who were funded. GOMA and Ocean Leadership wrote and executed contracts with institutions on behalf of GoMRI, ensuring the initiative followed best practices for an organization funding science.

American Institute of Biological Sciences

The American Institute of Biological Sciences (AIBS) was brought in as a party to GoMRI in order to further distance the Research Board from any perception of influence. In the earliest weeks of GoMRI, Colwell suggested that there should be an organization that supported the Research Board that was independent from GOMA and BP. Colwell believed that this would promote transparency and ethical decision-making on the part of the Research Board. While it was unusual for an organization that funded scientific research to have another organization handle its administrative costs, Colwell believed that there needed to be mechanisms in place to ensure the fair administration and distribution of funding from BP to the Research Board.

In the 1950s, AIBS was formed as a component of the National Academy of Sciences. It now consists of 115 member organizations relating to the life sciences, including scientific societies, museums, botanic gardens, and universities.[33] Its mission is "to promote the use of science to inform decision-making and advance biology for the benefit of science and society."[34] AIBS is a professional society that supports its membership in various ways while also advocating for and advising on sound educational, social, technological, and scientific policies in the life sciences. Given GoMRI's extensive focus on Gulf ecosystems and the effects of oil spills on the Gulf's flora and fauna, AIBS was a natural fit as one of GoMRI's earliest partners. Colwell had previously served as President of AIBS and was familiar with the organization.

Rather than guiding GoMRI science policy, the AIBS aided in assuring the public and scientific community of the initiative's independence. Along with other AIBS staff members, Jennifer Petitt provided Research Board support including overseeing its expenses, beginning in late 2010. As a 501(c)(3), nonpartisan, nonprofit organization, the Research Board believed that AIBS was well-suited to acting as its own financial agent, among other duties. Hence, AIBS came to support the board meetings, board committees, and organized travel logistics, meetings, and expenses for board members. This arrangement helped the board maintain its independence from BP and GOMA.[35] AIBS and Ocean Leadership also prepared press briefings for the Research Board.[36]

In a 2020 interview, Petitt emphasized that the distributed management system encouraged the staff of partner organizations to view themselves less as allegiant to their home office and more as members of the GoMRI team. She regarded the leadership team in particular as having helped dissolve boundaries that might otherwise have developed between employees of different organizations.[37]

Northern Gulf Institute

The Northern Gulf Institute (NGI) was one of the first partner organizations under the GoMRI umbrella. Led by Mississippi State University, the NGI—a NOAA Cooperative Institute—brought together six research institutions and NOAA to provide applied and comprehensive scientific research on the Gulf of Mexico.[38] NGI includes an education and outreach program that offers the public information about various Gulf-related issues. NGI works to improve NOAA's connections to academic institutions around the Gulf of Mexico.[39] Furthermore, NGI's funding structure allows it to place scientists into the field quickly because it does not hire employees as civil servants and does not follow the federal hiring process.[40] In some ways, the NGI model of supporting sustained scientific research and capacity building in the Gulf of Mexico through collaborative endeavors can be understood as a precursor to GoMRI.

In 2010 and 2011, GoMRI and NGI quickly developed both financial and professional connections within the framework of their new partnership. In order to send researchers into the field as quickly as possible and begin laying the foundation for later investigations, BP provided an initial $10 million in grants to the NGI for quick response by its member institutions and others in the Gulf. Rather than providing funds directly to researchers, BP sent grants to institutions with the ability to select researchers to respond to the blowout. Considering the NGI history of coordinating applied research in the Gulf, its composition of six research and academic institutions located in the Gulf region, and the infrastructure it had built to connect the public to the knowledge that it generated, the NGI was a suitable partner to participate in the initial scientific response to Deepwater Horizon.

NGI provided crucial support for GoMRI using its expertise in information systems design and management and science communication. The NGI developed and managed the GoMRI Research

Information System (RIS), its education website, and produced and vetted original content for the main GoMRI website, thereby providing communication and administrative support for the GoMRI Research Board, the board chair, and the chief scientist.

The RIS is an information system that tracked programmatic data for all GoMRI funded research projects, personnel, and publications. It allows the general public to access those data through a web portal. The RIS also includes the GoMRI synthesis and legacy products developed from the RFP process, discussed in chapter five. The RIS served as a resource for information dissemination and as an administrative oversight tool to track program metrics. Additionally, the RIS assisted in monitoring publications for compliance with attribution and data sharing requirements.

The education website houses educational resources developed through GoMRI funded activities. Serving both formal and informal education users, the site offers audio and video products, classroom materials, and stories about scientists, teachers, and students involved with GoMRI research.

The NGI produced a steady flow of articles for the GoMRI main website, providing a broad audience with easy-to-read accounts of fieldwork, summaries of selected peer-reviewed studies, and profiles of scientists and students. All contents were created in collaboration with researchers so that the scientific community and the public could trust the source for high-quality and accurate information. Articles hosted on the GoMRI website provided content for other GoMRI outreach efforts, such as social media channels, quarterly newsletters, biweekly eNews, and presentations.

Ocean Leadership

A significant partner with GoMRI, perhaps with the most sweeping mandate, was Ocean Leadership. It played a role in working with the Research Board RFP committees in the tasks of drafting RFPs, advertising, and administrative reviews. Ocean Leadership also had a lead role in designing and maintaining the GoMRI website. Its staff assisted in executing contracts, providing financial oversight, and ensuring consortia contract compliance. Employees of Ocean Leadership were also key to GoMRI synthesis and legacy efforts. They organized workshops and drafted publications related to synthesis activities. Ocean Leadership amplified GoMRI's research with its experience in outreach programming. The Research Board, aware of Ocean Leadership's experience in supporting scientific research, forged a partnership with the organization in mid-2010, which set the stage for Ocean Leadership's central role in so many GoMRI operations.

Ocean Leadership was formed from the combination of two preexisting organizations: the Joint Oceanographic Institutes (JOI) and the Consortium for Oceanographic Research and Education (CORE). The two organizations combined their efforts in 2007, representing many "of the leading public and private ocean research and education institutions, aquaria and industry with the mission to advance research, education and sound ocean policy. The organization manages or has managed a variety of ocean research and education programs, such as ocean drilling, ocean observing, ocean exploration, and ocean partnerships."[41]

The Research Board invited Ocean Leadership to administer the peer review process and manage certain grants, in order to "enhance scientific opportunities through post-grant collaborations" and to publicize the results.[42] Furthermore, Ocean Leadership was involved in GoMRI outreach efforts. It published the GoMRI newsletter, updated the GOMRI website, and developed annual outreach work plans with various stakeholders, such as the scientific community, the GoMRI community of researchers, federal, state, and local governments, industry, and the general public.[43]

While Ocean Leadership was first involved with GoMRI in late 2010, it publicly announced partnership with the initiative in a press release dated April 26, 2011. In that release, Ocean Leadership noted that it had been involved in bringing together the federal government, academics, and industry experts in order "to identify the science needs in the Gulf." The product of that symposium was a twenty-page summary.[44] By nature of its research, support, and outreach programs, Ocean Leadership had connections to a wide variety of stakeholders in the Gulf of Mexico, across the country, and internationally, helping to bring researchers into GoMRI and relay the new knowledge that GoMRI funding generated.

Sea Grant

Established in 1966, the National Sea Grant program created federal and state partnerships across thirty-four networks of colleges, universities, federal agencies, and laboratories to conduct research and outreach. As part of its outreach, Sea Grant developed instructional programs, demonstrations, publications, and other means of communicating among scientists, researchers, people who make a living on the water, and the general public about oceans and the Great Lakes in the United States. Through NOAA, the federal government in partnership with universities, colleges, and other research institutions have collaborated to coordinate research on the Gulf of Mexico and respond to environmental challenges. With the support of Dr. LaDon Swann, Director of Mississippi-Alabama Sea Grant, the Research Board forged a partnership with the four regional Sea Grant programs (Texas, Louisiana, Mississippi-Alabama, Florida). Sea Grant reached not just scientific researchers around the Gulf of Mexico, but also diverse stakeholders in the region. The state-level Sea Grant programs offered ways to reach people whose livelihoods depend on a healthy Gulf of Mexico. Their connections to the national Sea Grant network also meant that GoMRI research could reach broader audiences.

In 2014, Dr. Steve Sempier, outreach and deputy director for the Mississippi-Alabama Sea Grant Consortium, created a special working group within Sea Grant called the Gulf of Mexico Sea Grant Oil Spill Science Outreach Team (SGOSS). The team was comprised of employees from four independent Sea Grant College Programs

(Texas Sea Grant, Louisiana Sea Grant, Mississippi-Alabama Sea Grant Consortium, and Florida Sea Grant). Sempier led the SGOSS team and worked with the four Sea Grant college programs to recruit and hire individuals to partner with GoMRI. According to Sempier, SGOSS, and Sea Grant more generally, shares and translates information between the local community of lay audiences and scientific researchers in a five-step process:

a. Solicit information from the regional community to seek out the questions that locals have.
b. Comb through the literature and sees what it says.
c. Translate that knowledge for lay audiences.
d. Deliver and disseminate that knowledge through emails, in presentations, in seminars, and in workshops.
e. Continue the loop if lay audiences have more questions.[45]

Sea Grant employees sifted through an immense amount of peer-reviewed research generated by consortia and individual researchers in order to answer pressing economic and ecological questions locals in the Gulf region posed.

Sea Grant also made the concerns and questions of lay audiences in the Gulf Region known to research organizations. It had long been a trusted organization in the Gulf region and had proven over the course of nearly fifty years to be a reliable and knowledgeable organization. Thus, Sea Grant connected the Research Board, GoMRI-funded scientists, and the general public by conveying knowledge back and forth between researchers and lay audiences.

For Sea Grant, GoMRI offered an example of how to build interdisciplinary connections in the midst and aftermath of a disaster that cut across academic and geographic boundaries. Sempier and his team proved their ability to do extension, research, and education on the topic of oil spills in the Gulf of Mexico. He continued to be a superb partner after GoMRI funding expired, and results of the research are still delivered to stakeholders throughout the Gulf region. In an interview with the author, Sempier emphasized that there is a distinct need for regional organizations to be engaged in cross-disciplinary work such as Sea Grant performed in partnership with GoMRI.

Getting the Plane off the Ground: Building an Effective Research Organization Employing Distributed Management

The Research Board, with its leadership and management teams, built an effective distributed management system that responded to an environmental disaster in the Gulf of Mexico. Concerned by precious hours and days slipping by that could otherwise be used to test, monitor, and record conditions in the Gulf of Mexico, Colwell and the early GoMRI leadership team used professional connections successfully to bring various institutions and agencies under the initiative's umbrella. The management team that was in place by mid-2011 amplified the effort, reaching out to home institutions and securing them a place in the GoMRI management structure.

Ultimately, GoMRI was designed to grant partner organizations and research consortia oversight of their own day-to-day activities. While Wilson, Shaw, and Carron might step in to help guide consortia, small research teams, and individual investigators to ensure proper financial management and contract compliance, they intervened rarely during the decade of research that GoMRI supported. GoMRI thrived mainly because the institutions that it partnered with were respected authorities in the Gulf region with long and trusted legacies generating scientific knowledge and communicating that knowledge to relevant audiences. Under the direction of the Research Board, the GoMRI management team and its partner organizations helped build a large-scale scientific organization and avoid constructing a highly centralized administration. This granted GoMRI the flexibility to enhance its mission by periodically reassessing its role in disseminating new scientific knowledge to regional stakeholders. The board was able to rely on a bevy of talented managers to connect the public to the science it pursued.

In the final year of GoMRI funding, states and localities across the United States began to shut down in response to the COVID-19 pandemic. Although it interrupted valuable work in research facilities and laboratories on the Gulf Coast, the pandemic did not significantly disrupt the functioning of GoMRI's financial or administrative units.[46] Nor did the pandemic inhibit GoMRI management functions, beyond prompting various members of the management team to work with researchers and consortia to develop contingency plans. Born out of one crisis, GoMRI would complete its work in a second, albeit different crisis. Yet, because of its distributed management structure, GoMRI was well positioned to maintain standard operation without disruption. GoMRI provides a model of how a major organization, tasked with distribution of $500 million over ten years, can be built rapidly and effectively from the ground up.

CHAPTER THREE

OPEN, TRANSPARENT, AND ACCESSIBLE DATA
GRIIDC and GoMRI Data Management Policy

We quickly found that PIs in our GoMRI culture happily embraced their roles in working within GRIIDC's novel data archival system. This all quite simply worked very well (and was remarkably successful). I was not at first aware of how rare an occurrence this was until years later I accepted the role of Editor-in-Chief of the Journal of Geophysical Research Oceans. The bewilderment and recalcitrance of authors asked to deposit "their" data in some searchable form was astonishing. The entire GoMRI community set new high standards for skill combined with collegiality in creating this remarkable asset (the practice and expectation of including DOIs in publications has now spread to become a standard amongst AGU and other journals).
—DR. PETER BREWER, Monterey Bay Aquarium Research Institute[1]

From its earliest days, the initiative remained committed to making data open, easily accessible, transparent and available well into the future. This chapter traces the history of GoMRI's open data program, development of its data management policies, and initiative successes in capturing and providing data from researchers it funded. The history of the GoMRI data management policy is told in three parts. The first is how the Research Board and management team translated Master Research Agreement stipulations concerning data management into clear rules for data compliance. The second is the development of the Gulf of Mexico Research Initiative Information and Data Cooperative (GRIIDC), which served as a repository for all the data produced through GoMRI funding. The third part is how the board, leadership team, and other GoMRI managers ensured that researchers adhered to GoMRI's data policies.

Throughout 2010 and early 2011, the Research Board considered different options for how best to fulfill the MRA mandate about managing data and building a research database. Ultimately, GRIIDC became a foundational piece of the GoMRI data management program. It played a crucial role in GoMRI's pursuit of open, transparent, and accessible data about the Gulf of Mexico. GRIIDC was important in two major ways. First, GRIIDC allowed GoMRI to keep all of the data that its funding produced in a single repository. Second, GRIIDC underscored GoMRI's commitment to transparency and openness. Anyone with an internet connection may access GRIIDC. For nonscientists, GRIIDC showed that GoMRI was indeed putting the BP funding to good use by pursuing the best science, which produced abundant and openly accessible data. For researchers, GRIIDC has left a legacy of data collection that will help improve the state of knowledge about oil spills in and around the Gulf.

Turning Stipulations into Policy: The Language of the RFPs and MRAs and the Formation of a Comprehensive Data Policy

Colwell and the Research Board understood that data collection and management were crucial features of open, transparent, and accessible science. In 2010, however, it was not the norm in some fields for scientific data to be made readily available to the public, nor were many researchers aware of the best practices by which to facilitate data sharing.[2] Most researchers considered the data that they gathered to be proprietary information and keeping it from the public until published—and often after—was common. Releasing data too soon could jeopardize the scientist's ability to publish their research, as those data could be used by other parties without official acknowledgment. After publication, data generally remained proprietary, belonging either to the researcher or the institution that funded the scientific research; in some cases the data were stored somewhere on a shelf and later lost. While other organizations and agencies in the United States and internationally were beginning to move toward making data publicly accessible, what GoMRI offered was a multidisciplinary range of data that was expediently archived and made easily accessible to anyone with an internet connection.[3]

The reason for GoMRI's insistence on open data was linked to conditions in the Gulf region during the summer and fall of 2010. The Master Research Agreement included language pertaining to data accessibility. Transparency in all matters was one of the GoMRI foundational principles and it extended to data that scientists gathered using initiative funding. As mentioned in chapter one, the press and the general public were concerned at first that scientific researchers investigating the aftermath of Deepwater Horizon might not act with the community's interests in mind. Some of the scientific researchers and experts hired by BP initially underestimated the amount of oil that poured into the Gulf of Mexico between April and September 2010, which sowed doubts in the Gulf community about the trustworthiness of scientific experts. Gulf Coast residents sought scientific advocates whose impartial study of the spill's aftermath could guide restoration and rehabilitation. Colwell and the Research Board devised GoMRI to fill that role. By providing all the data, with complete transparency, GoMRI ensured its research database would help expand the reach of consortia and small team efforts including by providing basic information to all interested parties. The research database could also assuage concerns about scientific research being produced opaquely and out of the sight of the local community.

Although the first comprehensive MRA called for a research database, it did not explicitly mention GRIIDC. GRIIDC was the product of conversations after the MRA was in place. The earliest MRAs (versions 1 and 1.1) provided frameworks for how data management was to proceed within GoMRI, but it was up to the Research Board to interpret the text of the MRA and the management team to carry out their instructions. By studying the changing references to data collection, management, and publication in the GoMRI guiding documents, it is possible to depict the formation and evolution of the organization's data compliance policy over the course of the decade that it operated.

The revised version of the Master Research Agreement, dated July 11, 2012, gives additional insight into the ways that the research database would be developed and managed. Although the Research Board held sole authority to put forward policies pertaining to the database, the MRA instructed the board to consult with the GoMRI administrative unit and directors of the research consortia. The MRA encouraged collaboration among parties that would be involved in data management, rather than simply handing down instructions from above. Allowing representatives of the research consortia to participate in development of the database system would grant researchers, who were gathering the data, the ability to decide, in part, how those data would be collected, organized, and presented. The Research Board did not prescribe precisely how data were to be collected, only that the data were to be collected in a way that would allow it to be stored and later accessed by any interested party.[4]

The main team on the Research Board that helped draft and refine the GoMRI data management policies was the Data Committee. The Data Committee emerged early in GoMRI history, forming clear policies pertaining to data management and building on the MRA call for open and accessible data. Any consortium interested in acquiring or renewing funding commitments from the Gulf of Mexico Research Initiative first submitted a proposal in response to an RFP, as described in the previous chapter.[5] Language within the RFPs that pertained to data management changed as GoMRI circulated subsequent solicitations. RFP-I, dated April 25, 2011, required consortia to describe how they would furnish data to the Gulf of Mexico Research Initiative database and website.[6] RFP-I also stated that part of the GoMRI funding from BP was to establish and maintain "a publicly accessible, transparent database of information generated by the GRI [GoMRI] that is consistent with modern informatics [and] best practices."[7] GRIIDC, however, did not yet have a name.

In RFP-I, the Research Board made clear that "[t]he provision of a full and coherent database accessible to all scientists, policy makers, and the public is an essential outcome of the GRI."[8] From the moment that they applied for funding, PIs leading the research consortia were informed of their responsibility to furnish data to the GoMRI database. They were required to build the necessary infrastructure to capture, collect, and organize data produced from their research. Applicants' submissions were judged in part on their ability to manage data. Among the RFP-I evaluation criteria, 18 percent of the Research Board consideration of the proposals was given to the consortium management policies, including its data compliance plan.[9] From the outset, the GoMRI Research Board, represented by Dr. Peter Brewer, made it clear that its data management policy was important and applicants were expected to adhere to its interpretation of open and accessible data.

By 2013, the RFPs included a much more stringent data compliance policy. The RFP-IV evaluation criteria, which covered grants for the period spanning 2015–2017, allotted 15 percent of the proposal evaluation to management, which included data collection and organization. According to the language of the RFP-IV solicitation, successful applicants would build a "governance and communication structure proposed to support effective exchange of ideas, data and results among the sub-projects in the collaborating institutions to ensure cross-fertilization."[10] The point of that stipulation was to encourage broad collaboration and dissemination of information among the partner institutions and laboratories of the applicants. This fit with the MRA mandate that GoMRI conduct interdisciplinary research across many different campuses and institutions. Regarding data collection and management, RFP-IV required submissions to include strong "data management policies, favorably including a strong record of previous submission of data to a public database (where applicable, of GoMRI data to the Gulf of Mexico Research Initiative Information & Data Cooperative, GRIIDC)."[11] Research consortia were required to describe the type of data to be produced, standards for data and metadata format, policies for accessing and sharing data, policies and provisions for the reuse of data, and plans and timelines for archiving data.[12]

Clearly, the Research Board stipulations about data collection and management became more comprehensive and specific between the issuance of RFP-I and RFP-IV. For example, the data management section of RFP-IV references the fact that "all data should be made available with minimal time delay, through submission to the GoMRI data archive (GRIIDC) for the advancement of knowledge and utility to researchers, agencies and others."[13] A memorandum from the Research Board to RFP-I and RFP-II investigators released on December 2, 2013, had previously clarified what was meant by "minimal time delay," a phrase that had first appeared in the Master Research Agreement. The memorandum stipulated that "the Research Board policy interprets the meaning of 'timely manner' and 'minimum time delay' as no later than the date of publication of results. If a scientific publication is not expected within 12 months of the data being collected then the interpretation of 'timely manner' and 'minimum time delay' is no later than 12 months after the collection of the data."[14] The Research Board took care to ensure that GoMRI-funded consortia would follow this stricter data management policy for RFP-IV in subsequent rounds of funding solicitation. The memorandum notified consortia that any request for an extension in terms of data submission would be regarded as "an extraordinary exception" that "should be carefully justified."[15] The Research Board also noted that it would consider consortium compliance with the data management policy in future RFP submissions, extensions, and other funding opportunities.[16] The language of the RFPs became clearer and stricter about data compliance as the years went on.

Over the course of two years, from December 2011 until December 2013, the Research Board, the leadership team, GoMRI staff located at the Harte Research Institute (discussed below), and others worked to translate the spirit of the MRA language pertaining to data management into a clear and comprehensive data management policy. While the MRA provided a framework for the GoMRI data management policy, it was up to the management team to clarify and enforce the language of the MRA. Ultimately, the management team determined that the rapid collection and dissemination of data would best serve both the scientific community and the general public. Presenting the language of the MRA in terms of time limits, accessibility, and transparency clarified the Research Board's expectations for the research consortia, which would have to submit their own data management plans as part of their funding applications. In addition to the data management policy, however, during those early years the GoMRI Research Board also had to figure out how best to build a research database capable of handling information generated by the consortia and individual researchers.

Creating an Independent Data Cooperative

As noted above, the December 2011 version of the Master Research Agreement included references to the GoMRI research database. However, the language in that early MRA did not include specific references to GRIIDC. Much like GoMRI's formal articulation of its data management policy, it would take the first few months of the initiative's existence to build its research database system and to determine how it would help fulfill the data management framework laid out in the MRA. To help build a data storage system for GoMRI, the Research Board recruited help from the Harte Research Institute (HRI), located in Corpus Christi, Texas.

By intervening in the creation of GoMRI during the summer of 2010, the Gulf state governors had done much to focus the initiative on the region that had been most affected by Deepwater Horizon. The Research Board took its responsibility to build capacity in the Gulf and rely on institutions located there into consideration when it selected the Harte Research Institute at TAMU-CC to house the GoMRI research database. The Harte Research Institute had a great deal of prior experience working with scientific data management and monitoring in the Gulf of Mexico region.

The Harte Research Institute was founded in 2000, when local newspaper publisher Ed Harte (1922–2011) provided a $46 million grant to TAMU-CC to further research on the Gulf of Mexico.[17] Harte was highly regarded as a prominent conservationist in the region and the donation he made to TAMU-CC proved to be instrumental in building an organization capable of responding quickly to the Deepwater Horizon disaster. In 2001, Dr. John W. Tunnell Jr. was appointed associate director of the Harte Research Institute. Tunnell oversaw the development of HRI's research agenda and graduate program.[18]

Dr. Larry McKinney, who was serving as senior executive director of the Harte Research Institute at the time of the Deepwater Horizon disaster, was instrumental in connecting HRI to GoMRI. By 2010, McKinney had already earned a distinguished reputation in the discipline of oceanography and was known for his work in building research institutions focused on the Gulf of Mexico. For example, McKinney was a founding member and chair of the Gulf of Mexico University Research Collaborative. He also served on the policy committee of the Consortium for Ocean Leadership.[19] Apart from his work with the Harte Research Initiative, McKinney would go on to serve as the chair of the program committee for the GoMOSES conferences (discussed in detail in chapter four) in 2018, 2019, and 2020. Harte's collaboration with GoMRI on the creation and management of a large research database brought McKinney closer to the initiative. He would continue to play a key role within GoMRI through 2020.

In addition to McKinney, starting in July and August 2010, members of the Research Board reached out to Dr. James Gibeaut, who since September 2007 had held the position of Endowed Associate Research Professor in Geospatial Sciences at the Harte Research Institute.[20] By 2010, Gibeaut had already worked extensively on data collection and management.[21] He describes himself as "a coastal geologist who uses optical, radar, and lidar remote sensing, GIS, and field surveys to measure and understand coastal change."[22] The act of measuring coastal change using such a wide array of tools meant that Gibeaut produced a large amount of data in his work, which necessitated databases to store the information. Furthermore, Gibeaut played a role in the scientific community's response to the *Exxon Valdez* oil spill from July 1989 until July 1991. He traveled to Alaska soon after the spill and, during his two years there, had various jobs that required "collecting geological, chemical, and biological data at selected sites through winter, under difficult logistical and environmental conditions" and "[s]upervising employees and contractors responsible for collection and analysis of data concerning the Exxon Valdez oil spill."[23] Thus, Gibeaut had a great deal of experience with data collection and management during the *Exxon Valdez* oil spill, the most widely recognized oil spill until Deepwater Horizon, twenty-one years later.

In the summer of 2010, McKinney recruited Gibeaut to begin drafting a plan for building the GoMRI research database. By late 2010, Gibeaut and his team came up with the name GRIIDC, an acronym that stands for Gulf of Mexico Research Initiative Information and Data Cooperative. Initially, the final "C" stood for center, but Gibeaut decided that the name should reflect the collaborative nature of GoMRI. Hence, he decided the official name for the research database would be the Gulf of Mexico Research Initiative Information and Data Cooperative.[24] During the final months of 2010, Gibeaut and his team at the Harte Research Institute worked on assembling the storage capacity needed by GoMRI. With most of the equipment required to build GRIIDC in place by early 2011, the cooperative was officially established in 2011, when Gibeaut was offered a contract by GoMRI to handle the data its funding was generating.[25] The Research Board also put in place an Advisory Committee to steer GRIIDC. It included the board's Data Management Committee, CSO Chuck Wilson, a representative of GoMRI administration, designated data managers from each consortium, a representative from NOAA who specialized in data management, and Gibeaut as director.[26]

Like GoMRI, GRIIDC changed over time, but its core commitment to FAIR (findable, accessible, interoperable, and reusable) data principles remained consistent.[27] For one thing, GRIIDC is accessible to anyone with an internet connection. No login credentials or accounts are required to view the data GRIIDC stores. Given the voluminous data for which GRIIDC is responsible, much of the effort to adhere to FAIR principles has centered on improving the GRIIDC search engine. Readers are well aware that GoMRI collected data from several different fields in the natural and social sciences. To help navigate its many datasets, over the course of its operation GRIIDC has provided twelve publications that offer recommendations for searching through each of its datasets. Gibeaut and his team at Harte have also invested time and effort in improving GRIIDC's software. Since 2011, GRIIDC has adopted new software that has improved users' ability to sift through the immense data it stores. Initially, GRIIDC had limited search functionality; a query would only return a list of datasets relevant to the search terms. Successive software updates improved users' experience on the GRIIDC website. Now, as GRIIDC team members Rosalie

Rossi, Deborah LeBel, and Jim Gibeaut have written, users can "enter advanced search terms and narrow down to specific fields such as dataset title, abstract, author, or theme keywords. Facets can be used to further filter results by dataset status, funding organizations, and research groups."[28] GRIIDC continues to store data from multiple programs in the Gulf region. It has commemorated its connection to GoMRI by maintaining a subsite just for data generated with funding from GoMRI.[29]

GoMRI's Data Acquisition Policy

With GRIIDC largely in place by the summer of 2011, another set of questions emerged: how could the Research Board, the chief scientific officer, and the GRIIDC team ensure that researchers submitted data on time? What would be a reasonable amount of time in which to submit data? Who would be responsible for holding researchers to task? It was one thing to build a research database but creating a set of research policies was necessary to ensure proper operation of the database. Since furnishing data to an accessible database was far from standard policy in the early 2010s, how would GoMRI build and maintain a culture of open and transparent data among its many researchers?

In the actual wording of the research database policy, according to the earliest versions of the Master Research Agreement, the administrative unit was tasked with managing "(either directly or by means of a subcontract to a third party) a fully accessible database of results and ancillary information, such as metadata (the 'Research Database') and shall ensure that all data shall be posted with minimum time delay."[30] By the summer of 2011, the Research Board had decided to recruit a contractor to build the database—the Harte Research Initiative—but the rest of the language pertaining to that database was fairly ambiguous. Ultimately, as detailed above, the parties agreed that all data must be submitted within twelve months of the date of collection. This became the official interpretation of what the Master Research Agreement meant by "minimum time delay," although GRIIDC emphasized that researchers should try to submit data as soon as it was ready.

In late 2011, however, the Director of the Coastal Waters Consortium raised the question of timed datasets; that is, how would GRIIDC account for data acquired at different times as part of a series of investigations? GRIIDC responded by requiring timed series to be divided into one-year sets, ensuring that CWC would still adhere to the data management policy.[31] The team at GRIIDC emphasized timely data submission, and balked at the notion of researchers dragging their feet during the submission process, even for ongoing testing and monitoring.

The question of timing was not mundane but was crucial to the mission of the Gulf of Mexico Research Initiative. As noted in chapter one, many of the scientists involved in the formation of GoMRI had experience working with the *Exxon Valdez* disaster. In the wake of that spill, researchers found it necessary to submit data as quickly and efficiently as possible. If there were not clear and specific guidelines in place for submission of data, crucial information needed in future responses to major oil spills could be lost. The aftermath of the Ixtoc I explosion also informed the GoMRI approach to data management. Data collected from the *Exxon Valdez* spill was helpful, even if not as comprehensive as that collected after Deepwater Horizon, but the bulk of the data from the earliest days of Ixtoc I were absent due to technological standards of the day and the reluctance of the government of Mexico to allow investigations from US-based scientists. Ixtoc I, as discussed in the first chapter, was similar in nature to the Deepwater Horizon blowout and information from that disaster would have been helpful to first responders and policymakers planning the immediate and medium-term responses to the Deepwater Horizon 2010 spill.[32]

With the GoMRI data policies clarified and relayed to the research consortia between late 2011 and early 2012, a final question remained: how could GoMRI ensure that researchers would adhere to the data management policies? It was one thing to stipulate researchers submit their data in a timely fashion, but another to have them comply. Who should bear responsibility for keeping researchers on task regarding data submission? Initially it was unclear to the Research Board and Gibeaut if GRIIDC should play a role in data policy enforcement. The stakes were high. Between 2012 and 2014, some research consortia bristled under what they perceived to be strict data submission standards.[33] Yet, as noted above, time was essential to proper functioning of GRIIDC. The Research Board wanted data to be made publicly accessible as quickly as possible, within reason. This was in the interest of transparency and to assist in ongoing and future oil spill responses. Ultimately, Wilson stepped in to relieve GRIIDC of its responsibility to enforce the data policy, placing that responsibility primarily on himself and on Kevin Shaw, the program manager for GoMRI.[34]

Working with the Research Board, Wilson and Shaw determined that the best way to encourage data submission in a timely fashion was to tie data requirements to funding. The Research Board concurred. On December 2, 2013, the Research Board circulated a memo detailing the connection between funding and data submission to RFP-I and RFP-II recipients. In the second line of that memo, the board wrote "the Research Board will require all consortia and RFP-II individual projects to be in compliance with GoMRI data submission requirements prior to approval of 2015 continuation."[35] In a 2015 memo, Colwell emphasized the importance of timely data submission. She wrote that "the Board's, and your, efforts in the area of data sharing are thus not arbitrary, but are at the front of the curve in ensuring data transparency."[36] Colwell went on to write that continuation of funding would be contingent on consortia adhering to the data management policy of the Master Research Agreement.[37]

By 2017, the GoMRI Research Board had increased its expectations for data compliance. In the final round of GoMRI RFPs (RFP-VI, which covered the years 2018 and 2019), released on

October 3, 2016, the Research Board required data compliance training.[38] According to a memorandum from Chuck Wilson that was circulated to all RFP-VI consortia and individual researchers, "all RFP-VI Principal Investigators, co-Principal Investigators, and Data Managers (whether as members of Research Consortia or Individual Investigators or Collaborative Efforts) that have not done so must complete the 'Introduction to the GoMRI Data Management Program' training session no later than 90 days of the effective date of the grant agreement."[39] The training was split into two parts: "Organizing Data" and "How to Submit Data to GRIIDC."[40] The data managers, PIs, and co-PIs could receive their certification of compliance by participating in webinars, attending workshops, or scheduling a one-on-one meeting with the GRIIDC program manager. The Research Board's decision to require data compliance certification indicates how seriously it took its data management policies. GoMRI funded research would be held to very high standards of data collection, management, and organization in the hope of improving responses to future Gulf oil spills.

The Research Board's decision to tie funding to adherence to the data management policy accomplished three goals. First, both Shaw and Gibeaut were able to report that the data management compliance rate was close to 95 percent.[41] Second, the compliance rate shows that it is possible for large scientific organizations to acquire and make data accessible to the public in a timely fashion. As noted above, this was significant improvement over acquisition rates during and after previous oil spills. Finally, GoMRI's adherence to timely data acquisition policies and transparency helped solidify its reputation as an organization committed to public advocacy and support. By making its data available in a timely fashion, GoMRI helped keep the general public and local institutions and agencies apprised of new information about the health and welfare of the Gulf and its response to Deepwater Horizon.

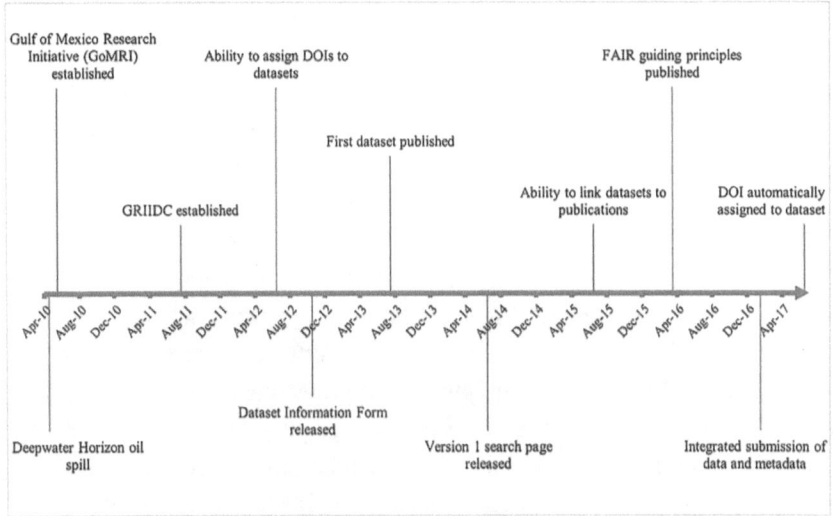

Figure 3.1. Timeline showing when major events in the history of GRIIDC occurred. Available in Rosalie Rossi, Deborah A. LeBel, and James Gibeaut, "Growing Pains of a Data Repository: GRIIDC's Evolution from Environmental Disaster Rapid Response to Promoting FAIR Data," *Frontiers in Climate* 24, no. 4 (August 2022): 1–6; https://doi.org/10.3389/fclim.2022.958533. © 2022 Rossi, LeBel and Gibeaut. Licensed under CC BY 4.0, https://creativecommons.org/licenses/by/4.0/.

Conclusion

GRIIDC has remained an enduring feature of GoMRI's research program. Five years after GoMRI ceased operations, GRIIDC continues to provide data to researchers, institutions, NGOs, and government agencies. Furthermore, between 2010 and 2020, BP transferred all of its data pertaining to Deepwater Horizon to GRIIDC for storage, archiving, and dissemination. Thus, GRIIDC remains an institutional legacy of the GoMRI support for scientific research. Even after the end of GoMRI, GRIIDC still serves as a vital resource for researchers, journalists, and the public. In addition to accepting new data, Gibeaut and the rest of the team at the Harte Research Initiative have focused on securing the database, given the Gulf region's propensity for storms and natural disasters. In 2017, it had its first test when Hurricane Harvey bore down on southern Texas. During that storm, GRIIDC was put into a read-only mode and was also backed up onto mirror servers at Texas A&M University–College Station. At no point did Harvey threaten public access to the GoMRI data.[42]

The GoMRI data management policies represented an important development in the history of scientific data management for three major reasons. First, the Research Board was committed to open and accessible data presented in a timely fashion. As noted in the first chapter, in the wake of Deepwater Horizon, the Gulf Coast community sought reliable scientific allies. By making all data publicly available, GoMRI thereby was committed to transparency, as well as collaboration in its approach to protecting the health of the Gulf community. The Master Research Agreement between BP and the Gulf of Mexico Alliance was more than just a means for scientists to achieve independence—namely to distance themselves from the party found responsible for Deepwater Horizon—it was also a framework for honest, open, and comprehensive research done on behalf of the Gulf community. GoMRI's data compliance policy is a vehicle for science in the public interest.

Second, according to Chuck Wilson, GoMRI was a pioneer in the way it helped change the mindset of the scientific community on acquiring and releasing data. The scientific community in 2010 was accustomed to treating its data as proprietary information and funding agencies did not routinely insist data be made publicly available within one year of publication.[43] As Rossi, LeBel, and Gibeaut have written, "[d]ata sharing was not as prevalent as it is currently, and many researchers were not familiar with data sharing and data organization best practices."[44] While this was not a universal truth—the genomics community and physical oceanographic community were familiar with sharing data openly and transparently—it was the case for some sciences. According to Wilson, at the time that the Research Board began building GRIIDC it was not common practice for the community of biologists to share data widely and with nonscientists. GoMRI offered a model for how to capture, store, and widely disseminate scientific data to any interested party.

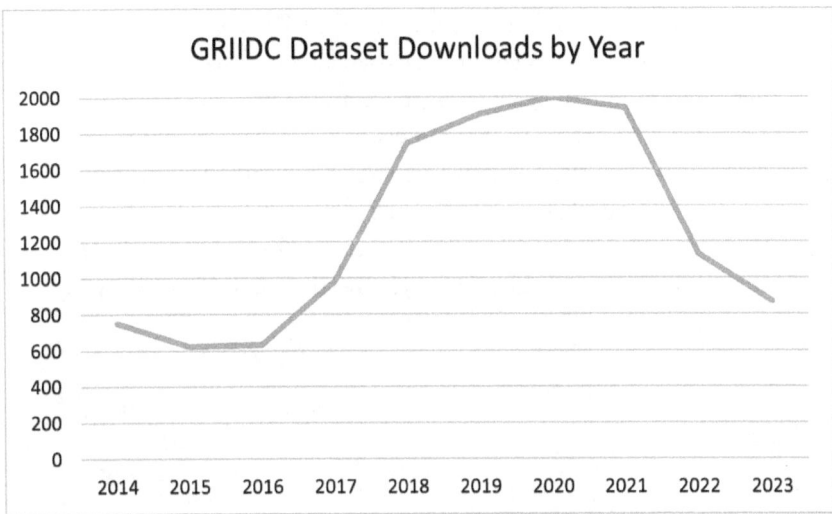

Figure 3.2. A figure showing the annual number of dataset downloads from GRIIDC.

Third, by using the newest tools available, and building its own comprehensive database, GoMRI demonstrated that voluminous data sources could be made available expeditiously to scientists and the public. NOAA, NHS, and NSF offer several different databases, each covering a particular scientific theme or area. GRIIDC is a single, unified database. A clear benefit of this approach is that scientists do not need to do preliminary searches to determine where data relevant to their work are located. Although GoMRI employed NHS and NSF databases as models, Colwell, very familiar with the issues of data management, chose to improve on what already existed. As the former director of NSF, Colwell was very familiar with the benefits and drawbacks of existing models and tasked the relevant individuals within GoMRI to build an improved and more effective database for the Gulf of Mexico.

GRIIDC will prove to be an enduring legacy of GoMRI. It continues to accept data from several Gulf-based institutions.[45] To mark its legacy, the data that were gathered using GoMRI funding have since been preserved as a subset of GRIIDC's larger collection, showcasing what GoMRI was able to accomplish. GRIIDC also remains relevant to researchers today. In 2023, users downloaded 869 datasets that comprised 625 gigabytes of space. In the first quarter of 2024 alone, users downloaded 465 datasets, or more than half of the total downloads from the previous year.[46] As it continues to accept data from institutions that conduct research across GoMRI's five themes, GRIIDC will remain an important part of efforts to better understand the Gulf of Mexico. Through GRIIDC, GoMRI's legacy of scientific research lives on.

CHAPTER FOUR

OUTREACH AND CAPACITY

Building a Legacy of Oil Spill Science in the Gulf of Mexico

The Research Board was committed to making sure that the research it funded would not only shape future oil spill response strategies but also deepen public understanding of the Gulf of Mexico and its ecosystems. To achieve this, we made it a requirement for grantees to carry out meaningful outreach that engaged all stakeholders, including the public. We also wanted to highlight results across the entire program, so we brought in national experts like Sea Grant, the Smithsonian Institution, and Screenscope Productions to enhance and amplify the researchers' outreach efforts. These efforts made outreach a valued aspect of the GoMRI research culture.

—DR. MARGARET LEINEN, GoMRI Research Board Vice-Chair

One year after the Deepwater Horizon disaster, Harte Research Institute Executive Director Larry McKinney wrote that he "was filled with dismay and frustration upon hearing a senior federal agency administrator—charged with managing our nation's ocean waters—write off the Gulf before an audience of agency staffers."[1]

He went on to note that "I have stood silently seething as a leader of a major conservation group stated that there was nothing worth saving in the Gulf, that it was a lost cause."[2] Some scientists working in the region have offhandedly referred to the Gulf of Mexico as a forgotten sea. Despite its size and significance to the economy of the United States, federal expenditures for research in the Gulf of Mexico have been less, on average, than for other bodies of water.[3] A GOMA white paper published in 2013, two years into GoMRI operations, shows that EPA projected expenditures for the Gulf of Mexico totaled only $4.4 million, which was far less than the amount allocated to the Chesapeake Bay ($72.6 million) and the Great Lakes ($300 million) in that year.[4] Those amounts were obviously not correlated with the size of the bodies of water and represented a concern expressed by a handful of scientists in the Gulf region: that the federal government simply did not wish to spend an appropriate amount of money on research in the Gulf of Mexico. As the chair of the conference organizing committee, McKinney was asked to deliver a speech during the final GoMOSES conference in 2020. Toward the end of his remarks, he reiterated what had become a common refrain in his publications. McKinney again referred to the Gulf of Mexico as America's forgotten sea, receiving less attention from the federal government than other major bodies of water. He concluded by stating that only the Gulf of Mexico community was going to put up an appropriate amount of money to fund research and investigations of the Gulf of Mexico, as history has shown time and time again.[5] In his remarks McKinney expressed a sentiment shared among some of his colleagues in the Gulf region. Although this is not a universal sentiment shared by the community of scientists involved with GoMRI, McKinney's stature in the Gulf research community gives him an important platform to advocate for increased research funding in the Gulf of Mexico region.

This chapter describes some of the ways that the GoMRI worked to enhance the scientific capacity of the Gulf region, educate the public about Gulf issues, and connect scientific research to user communities. With $500 million targeted to support a scientific research program on the consequences of the Deepwater Horizon oil spill, GoMRI was well-positioned to bring more attention to and resources for scientific research in the Gulf Coast region. Before the first call for proposals was released, the Research Board had concluded that GoMRI funding of the best possible scientific research would include connecting knowledge generated to stakeholders. Gulf communities could benefit from enhanced understanding of the Gulf of Mexico. With the intent that the science supported should benefit the region, GoMRI made its decision to help Gulf Coast communities understand the impact of oil and oil spills on the marine environment and provide a knowledge foundation for restoration and protection.

GoMRI conducted outreach efforts to engage various stakeholders in the Gulf region and connect them to the scientific research being done. Because GoMRI relied on institutions in the Gulf region and around the world to perform interdisciplinary research, it was well-positioned to conduct outreach on a variety of topic areas across the entire Gulf Coast. Research consortia and the GoMRI management team developed outreach programs to communicate to diverse stakeholders with varied interests the scientific knowledge generated. To do so, GoMRI first identified relevant stakeholders and mechanisms by which to understand and address their concerns and questions. The GoMRI outreach program was an integral part of its scientific mission and, thereby, the knowledge gained remains accessible in various forms long after the initiative ended.

During the initial conversations about how to build GoMRI, BP expressed a desire to build and improve scientific research capacity in the Gulf region. This became a guiding principle in the formation of GoMRI. The Research Board developed several methods to enhance research capacity in the region. It is important to note that the research capacity was not limited to oil pollution alone; GoMRI's capacity-building efforts were designed to encompass environmental quality and curiosity-driven research. GoMRI reliance on interdisciplinary research done by institutional teams forged new partnerships and connections that carried forward after 2020. Supporting early-career scientists and graduate students was an extremely important component of the initiative, notably to enhance the scientific research capacity of Gulf Coast institutions.

Education was also an important part of GoMRI's outreach and capacity-building efforts. GoMRI-funded consortia used outreach to develop curricula for K–12 education. GoMRI also partnered with organizations with expertise in educational publication and programming to connect broader audiences to the science.

GoMRI Consortia Outreach Activities

From the first days of the initiative, outreach was a crucial component of the GoMRI model. Not only was it specified in the Master Research Agreement, but Colwell and the entire Research Board knew they needed to seek, support, and guide the best scientific research and communicate it to the public. The consortia became the main drivers of outreach efforts, with support from the Research Board Outreach Committee along with the management team. Because each consortium assigned its own outreach coordinator—a requirement stipulated in the RFP—a number of unique outreach programs were created. The Research Board assessed the outreach strategies of the consortia and offered suggestions. However, the consortia outreach programs were administered not by the GoMRI board but by the individual consortia.

The history of the Center for the Integrated Modeling and Analysis of the Gulf Ecosystem (C-IMAGE) offers a helpful anatomy of a successful consortium outreach program.[6] C-IMAGE comprised nineteen partners located in the United States and around the world. Its research addressed the GoMRI themes of chemical evolution and biological degradation of petroleum and dispersant mixtures and of environmental effects of petroleum and dispersants on the Gulf ecosystem.[7] C-IMAGE received funding by RFP-I, RFP-IV, and RFP-VI and, therefore, was a key component of the GoMRI scientific program since 2011. C-IMAGE was led by Chief Investigator Dr. Steven Murawski and Chief Scientist Dr. David Hollander.[8] Dr. Teresa Greely led C-IMAGE outreach.

C-IMAGE conducted outreach in several different ways. Rather than relying on a single program, C-IMAGE outreach staff, in collaboration with the GoMRI management team, created a variety of media and programs to connect wider audiences to the scientific research. For example, the consortium produced a series of podcasts covering several topics, including examples of the objectives of C-IMAGE research, how oil affects the Gulf ecosystem, and potential future oil spill response scenarios. One of the podcasts discussed the scientific community response to Deepwater Horizon, comparing it to that of the 1979 Ixtoc I oil spill in the Southern Gulf of Mexico. Titled *The Loop*, the podcasts were produced by Mind Open Media, with transcripts in both English and Spanish, thereby reaching a wider audience.[9]

In addition to the podcasts, C-IMAGE consortium also provided educational opportunities and materials to teachers in the Gulf region. Two such programs merit mention here. The Teacher @ Sea program offered K–12 educators the opportunity to participate in scientific expeditions in the Gulf of Mexico. While onboard, educators became members of the scientific crew, acting as "ship to shore communicators." In that role, teachers recorded videos about their experiences on the research cruise, which were released on social media and in live classroom video chats.[10] The teachers also joined "marine scientists, post-docs, and graduate students participating in the C-IMAGE cruises" to "communicate the science, technology and life at sea experiences to educational and general audiences."[11] The Teacher @ Sea program was a significant outreach strategy that connected students in their classrooms to the scientific research that GoMRI funding supported.

Back on shore, teachers organized professional development programs, translating Teacher @ Sea experience into lesson plans that were integrated into regional elementary, middle, and high school science programs. Through this outreach program, C-IMAGE was able to reach students, teachers, and university researchers and connect educators and students to field work funded by GoMRI. The Research Board's decision to require consortia outreach and to evaluate research proposals based, in part, on outreach efforts played a crucial role in connecting a broad segment of the regional population to the scientific research done in the Gulf of Mexico.

The consortium outreach strategy had the advantage of allowing consortia to develop their own programs as deemed appropriate. This provided consortia flexibility in responding to the needs of the local community. Consortia awarded continuing funding in subsequent RFPs were able to build on their previous outreach programs and expand their audience. The C-IMAGE Teacher @ Sea program developed over several years and became a very successful means to link GoMRI-funded research to the local Gulf Coast communities. In keeping with GoMRI distributed management, the consortium outreach model accorded significant responsibility for oversight on consortium leaders and outreach directors who were strongly linked to the scientific research, and thus were best suited to understand needs, concerns, and questions of the local communities.

Research Board-Directed Outreach Activities

In its formative years, GoMRI was faced with questions about how best to build an outreach program. GoMRI encountered a Gulf community that was very diverse, comprising a variety of social groups with a complex set of relationships to the water and land around the Gulf of Mexico. Fishers, tourism industry, and oil companies, among others, each had a different understanding of the value of the

Gulf water and shore. GoMRI tailored outreach strategies to answer various questions these different groups—and others—posed.

To begin addressing such questions, in 2011 the Research Board established an Outreach and Communication Committee (OCC), chaired by Debra Benoit, to develop outreach strategies, carry out outreach activities, and review criteria of the RFPs for evaluating consortia outreach programs. To develop a completely new comprehensive outreach program, the committee addressed several questions in its first months, including: "Who are the stakeholders/target audiences? Which stakeholders can we most effectively inform? How do we determine and prioritize outreach messages? What should be the size and scale of the effort and can GoMRI afford it? What partners do we need to be maximally effective?"[12] The scope and scale of the outreach effort was not clear at first, but those questions would ground future outreach programs with a clear set of priorities pertaining to audiences, messaging, and partnerships.

The OCC enlisted the Consortium for Ocean Leadership to assist in drafting the general outreach plan. Ocean Leadership had a long track record of connecting different audiences to oceanographic research through outreach. Broadly, the outreach plan developed by the OCC, Research Board, and Ocean Leadership filled gaps in various stakeholders' knowledge about risk, consequence, and mitigation possibilities for coastal oil spills. The outreach plan was designed to reach diverse stakeholders by "developing content for traditional science media outlets that typically reach a relatively small audience of knowledgeable science enthusiasts" and "sharing oil spill related discoveries through popular science outlets such as documentaries, science magazines, graphic novel adaptations, podcasts, newspaper science coverage, science websites, and social media."[13]

The first challenging question that the OCC had to answer was who, exactly, were the stakeholders in the Gulf region? The OCC assumed different groups would have different questions for researchers based on the ways oil spills affected their livelihoods. The outreach team classified potential Gulf stakeholders into seven groups: society (emphasizing the general public), scientific community; GoMRI community; K–12 students, undergraduates, graduate students, and educators at all levels; government officials and policymakers; industry; nongovernmental organizations; and public health officials.[14] It would be the task of the OCC to connect those stakeholders to the GoMRI community of researchers and translate the knowledge produced to the diverse audience of the Gulf community.

Many GoMRI outreach efforts took place under the auspices of organizations independent of the initiative and with previous experience in education and outreach. As described in chapter two, the OCC contracted Sea Grant to play a key role in GoMRI outreach. The Sea Grant webpage offers a succinct summary of the organization that explains their mission and relevance to GoMRI outreach:

> The National Sea Grant College program was established by the U.S. Congress in 1966 and works to create and maintain a healthy coastal environment and economy. The Sea Grant network consists of a federal/university partnership between the National Oceanic and Atmospheric Administration (NOAA) and 34 university-based programs in every coastal and Great Lakes state, Puerto Rico, and Guam. The network draws on the expertise of more than 3,000 scientists, engineers, public outreach experts, educators and students to help citizens better understand, conserve and utilize America's coastal resources.[15]

Together, Sea Grant and OCC enhanced the GoMRI ability to conduct outreach efforts. Using those partnerships allowed the GoMRI management team to build an extensive in-house outreach program. The consortia could certainly do much on their own to engage stakeholders in the Gulf region, but outreach efforts directed from within the management team amplified the scientific work that GoMRI funded and reached national audiences.

The Research Board OCC implemented a routine to connect outreach managers located in the Gulf Coast. This gave the management team a better understanding of the different outreach programs taking place across the organization. In the early years of GoMRI, the entire management team participated in a weekly conference call. Once a month, the OCC, some members of the Research Board, and the outreach team managers took part in an hourlong conference call to discuss consortia outreach. Every other month the consortia outreach coordinators joined a conference call with GoMRI's outreach management team to provide updates. Every fall, the GoMRI management team participated in a strategy retreat, during which it drafted the annual outreach and communications work plan.[16] With outreach managers working with consortia located across the five Gulf states and a Research Board and management team that was also dispersed, distributed management kept the GoMRI management team informed and engaged with consortia outreach efforts.

Once the annual work plan was approved by OCC, the GoMRI management team used the organization's official website as an information clearinghouse to inform the public about different scientific and outreach programs that GoMRI supported. Consisting of articles written by Maggie Dannreuther of the Northern Gulf Institute and maintained by Ocean Leadership, the GoMRI website described the research undertaken by various GoMRI funding recipients and graduate students. The website also provided a history of the initiative, its leadership, and links to partner organization websites.

GoMRI also used print and virtual means of disseminating information to the general public and to maintain continuity of communication within the organization. It published four newsletters a year and sent them via email to over 2500 recipients.[17] The newsletters discussed information available on the GoMRI website. They have been archived on the GoMRI website since the first issue, which was released in winter of 2013.

Connecting GoMRI Science to Relevant Stakeholders

GoMRI relied on three partner organizations to extend outreach as broadly as possible. The OCC recruited partner organizations that had experience translating scientific knowledge to nonscientific audiences. Partner organizations drew in larger audiences. The three major partners that the Research Board recruited to conduct nonconsortia outreach efforts were the Sea Grant program, the Smithsonian Ocean Portal, and Screenscope Productions.

Sea Grant and Outreach

As described in chapter two, in 2013 GoMRI and the four Gulf Coast Sea Grant college programs formed a new partnership to share GoMRI research on the Deepwater Horizon spill and its impacts on the Gulf of Mexico ecosystem. The separate programs had extensive prior experience connecting the science generated by oil spill research to the local Gulf Coast. They had first done so in the aftermath of the Hurricane Katrina disaster, and periodically undertook joint topically relevant outreach and education efforts. GoMRI provided another opportunity to bring the regional Sea Grant programs together in response to the Deepwater Horizon oil spill. That partnership gave rise to the Gulf of Mexico Sea Grant Oil Spill Science and Outreach Program (SGOSSP).[18]

To assist with outreach efforts, among other responsibilities in its partnership with GoMRI, the SGOSSP recruited four extension professionals with advanced degrees in science disciplines. One extension professional was placed in each Gulf state Sea Grant. Each extension professional had experience researching at least one of the five GoMRI scientific themes. The members of the SGOSSP engaged with a wide variety of audiences, drawing on the knowledge, experience, and resources of the entire Sea Grant organization.[19] By partnering with GoMRI, Sea Grant researchers were able to bring together the knowledge generated from both GoMRI and federal research efforts. This gave the SGOSSP more resources to connect wider regional audiences to new research on the Gulf of

Mexico. Usually, Sea Grant outreach efforts took place within particular states and occasionally in multistate regions. Combining the efforts and staff of four state programs (Mississippi and Alabama share a Sea Grant program) overcame the political boundaries that sometimes constrained the reach of Sea Grant outreach efforts.

During its first two years, the SGOSSP produced several outreach publications that reflected the needs and interests of the seven target audiences described by the OCC.[20] Sea Grant was able to pinpoint various questions that communities asked about the Gulf region. Through social network analyses, Sea Grant assembled a list of 449 oil spill-related questions that GoMRI-funded consortia and small teams could answer. Sea Grant officials also met with 530 individuals across the Gulf region.[21] After figuring out what questions the Gulf community was interested in and who would most reliably disseminate information, the SGOSSP wrote several targeted publications about topics such as fishery management, seafood safety, and the tourism industry. The team added a science communicator to assist with developing and sharing the outreach products. The SGOSSP also produced internet videos on various topics and organized science workshops designed to bring scientists into conversation with local communities.[22] GoMRI's partnership with Sea Grant connected it to a network of thirty-four such programs across the country, allowing it to assist in informing the public about consequences of oil spills in various regions, including Georgia and California.

Smithsonian

In 2010, the Smithsonian Institution launched the Smithsonian Ocean Portal website in conjunction with the opening of the Sant Ocean Hall at the American Museum of Natural History in Washington, DC. On the occasion of the Ocean Hall opening, the Smithsonian featured prominent presentations, exhibits, and virtual spaces dedicated to ocean science. The Smithsonian referred to this range of ocean-focused programming as the Smithsonian

Ocean Initiative. Soon after the Deepwater Horizon oil spill and coincidentally just a few weeks after launch of the Smithsonian Ocean Initiative, the Smithsonian began featuring information about the spill on the Ocean Portal. The Smithsonian Ocean Portal quickly became another part of GoMRI outreach efforts. The Smithsonian amplified the work done by GoMRI scientists, while cross-promoting its Ocean Portal.

The official partnership between GoMRI and the Smithsonian dates to 2013, the same year that Sea Grant was contracted to participate in GoMRI outreach efforts. Like Sea Grant, the Ocean Portal brought GoMRI scientists in contact with professionals who had experience creating interesting and informative publications and websites based on the latest scientific research. The Ocean Portal provided summaries of research GoMRI sponsored in a variety of audio and visual formats. It also provided a platform that increased the ability of GoMRI to reach the public and communicate its research program and findings. Ocean Portal also expanded the social media reach of the initiative. By presenting the scientific research that GoMRI supported to thousands of social media followers, the Ocean Portal reached wider audiences.[23] More specifically, the Ocean Portal connected GoMRI-funded research to younger audiences. In 2021, one year after GoMRI ceased operations, the Gulf Oil Spill section was the most-visited page on the Smithsonian Ocean Portal, receiving over 200,000 page views that year.[24]

Screenscope

In 2014, through the efforts of the OCC, GoMRI circulated a request for qualifications (RFQ) to production companies to create a documentary based on research projects that the initiative funded.[25] The GoMRI management team did not have experience producing documentary films, so the Research Board concluded it appropriate to reach out to production companies. Members of the Research Board believed this would be a mutually beneficial arrangement, as GoMRI could then broadcast oil spill science in a way that audiences would find compelling and interesting and the independent filmmaker would be able to "develop and produce their personal artistic vision."[26]

As with the consortia, the selection process for the production company was the result of a competitive process. The Research Board required applicants to submit capability statements providing insight into their creative processes, track records, and ability to produce a high-quality documentary. Several applicants were considered during the selection process. Ultimately, the GoMRI Research Board selected Hal and Marilyn Weiner, filmmakers and producers located in Washington, DC.

The Weiners are the founders of Screenscope, a film production company specializing in documentary features. They have long been part of the Capital arts and culture scene. Marilyn Weiner served as the District of Columbia Commissioner for the Arts and Humanities for six years in the late 1990s and early 2000s. Hal Weiner is involved with various civic causes and has given lectures about his experience as a filmmaker at colleges and universities throughout the Washington metropolitan region.

During the RFQ process, the Weiners proposed their film accomplish four goals. First, the film would show how GoMRI-funded research was being used to restore the health of the Gulf of Mexico. Second, the film would prominently feature GoMRI funded researchers, showing how different scientists worked together to understand and respond to the Deepwater Horizon oil spill. Third, fulfilling part of GoMRI's commitment to capacity building and education, the series would inspire young viewers to be engaged with oil spill science. Finally, the film would make a strong case for the role of science in addressing contemporary social and environmental issues.[27]

The team titled their project *Dispatches from the Gulf* and decided to make a three-part documentary series showing the development of GoMRI funded oil spill science. *Dispatches from the Gulf I* was released on February 4, 2016, at the annual GoMOSES conference in Tampa, Florida. GoMOSES attendees the premiere of the hourlong documentary before it was distributed for public screening at museums and science centers throughout the Gulf Coast and around the world. After it was shown on regionally affiliated PBS stations in the Gulf of Mexico area, the film was entered for competition at the Washington, DC, Environmental Film Festival. *Dispatches from the Gulf I* was recognized by the filmmaking community with an Emmy for photography in 2016. Matt Damon's role as narrator lent a familiar voice to the film that appealed to mainstream audiences.[28]

The initial release was followed by two sequels, *Dispatches II* (2018) and *Dispatches III* (2020). Like the first film in the series, the sequels premiered at GoMOSES before being broadcast on other platforms. Along with each release came several short videos featuring researchers and graduate students who participated in GoMRI-sponsored scientific investigations. The videos were also distributed through social media to connect the general public to the different types of research projects that GoMRI supported. The Screenscope production team also published educator guides, lesson plans, and free visual material to build curricula and educational programs based on GoMRI-sponsored research.[29]

Building Capacity and a Lasting Legacy: GoMRI's Efforts to Encourage Oil Spill Research in the Gulf Region

Many scientists have recognized the Gulf of Mexico for its ecological diversity and economic value. Yet that recognition has not necessarily translated to equitable research funding from the federal government and other sources. The Deepwater Horizon oil spill offered an opportunity to focus research efforts on the Gulf of Mexico. Funding for scientific research in the Gulf came from a variety of sources. In addition to BP's contribution of $500 million to establish the Gulf of Mexico Research Initiative, 80 percent of the Clean Water

Act penalties that were levied against the company and other liable parties went to the Gulf Coast Ecosystem Restoration Trust Fund. In total the Restoration Trust Fund was provided with about $5.3 billion from the Clean Water Act's stipulations. The funding has been divided among states, the federal government (primarily NOAA), and research centers to support a variety of programs, including economic rehabilitation, Gulf Coast restoration activities, monitoring in the Gulf itself, and applied research.[30] Still, the Research Board believed that more needed to be done to build scientific capacity in the Gulf region. It did so through several different efforts.

The Consortium Model

The Master Research Agreement states that the consortia "should utilize Research Investigators recruited from multiple institutions chosen to provide complementary research capabilities with the greatest strength."[31] Although the MRA does not mention scientific capacity explicitly, GoMRI was established to fund the best science, and in the interpretations of BP, GOMA, and the Research Board, the interdisciplinary and cross-institutional connections forged by the consortium model were the best vehicles to achieve the desired outcome. The consortia increased scientific capacity in the Gulf Coast by conducting extensive scientific research, bringing the faculty and students of different universities into collaborations, fostering networks between research centers both within and outside of the Gulf region, and building outreach and education programs that would encourage sustained interest in the Gulf of Mexico.

The consortia provided one type of model for the capacity-building work that could take place in the Gulf region after GoMRI completed its mission in 2020. For example, the consortia comprised multiple research institutions that collaborated on one or more of the GoMRI theme areas. During a consortium operational lifespan, scientists and researchers from different disciplinary backgrounds worked closely on various topics. This meant that the consortia facilitated interdisciplinary professional relationships benefitting future research efforts. The consortium model also overcame the arbitrary nature of state boundaries in the Gulf region. The problems that one Gulf state faces in responding to a major oil spill are often experienced by other states. GoMRI built networks across state lines focused on applied oil spill science, which benefitted the region.

GoMRI funding offered more opportunities for graduate students to work under the supervision of scientists, enhancing the time that they spent working toward their degrees. At the closing of the program, GoMRI had supported more than 1,200 graduate students and 1,000 undergraduates. By increasing the number of active laboratories in the Gulf region, more graduate students were able to do laboratory work while writing their theses and dissertations. Some early-career researchers established new laboratories focusing on oil spill research in the Gulf of Mexico. Working within the five GoMRI scientific themes, the consortium model granted graduate students and early-career scientists more opportunities to conduct research on the Gulf of Mexico, thereby fulfilling the GoMRI commitment to enhance the general scientific capacity in the Gulf region.

The GoMRI Scholars Program

The GoMRI Scholars Program was an important part of the initiative's capacity-building efforts. A subset of graduate students working with GoMRI funded research teams were eligible to become GoMRI scholars. Candidates had to be graduate students who participated in a project, primarily funded by GoMRI, for at least one year. Candidates also had to base their eventual thesis or dissertation on GoMRI-funded research in which they had participated. The candidates were nominated by the PI of a GoMRI-funded research team. Once the GoMRI management team approved the candidates, they received a letter of congratulations and a certificate bearing Dr. Colwell's signature. GoMRI, in partnership with the NGI, created a database named the Research Information System to track the progress of GoMRI scholars (along with all other scientific publications based on research funded by GoMRI) by the publications that they produced during the first few years of their careers.

GoMRI built capacity by encouraging networking among researchers who received its funding. New GoMRI scholars were recognized annually at the GoMOSES conference and invited to a luncheon and seated with members of the Research Board. At the luncheons, the scholars mingled with fellow graduate students and members of the Research Board. GoMRI scholars were invited to give a brief talk at the conference to describe their experience working on GoMRI funded scientific projects. Ultimately, GoMRI was successful in supporting 326 GoMRI scholars, 637 PhD students, and 567 masters students.[32]

Gulf of Mexico Oil Spill and Ecosystem Science Conference

The Master Research Agreement between GOMA and BP stipulates that GoMRI awardees—both the consortia and small-team investigators—convene each year at an "annual meeting." In the first few years, the Research Board decided that its own annual meeting was a good opportunity to bring oil spill scientists and other stakeholders together to present new research and discuss the state of the field. Later renamed the Gulf of Mexico Oil Spill and Ecosystem Science Conference (GoMOSES), the annual meeting would come to host hundreds of professional scientists and students engaged in oil spill research.

GoMOSES began as the annual meeting of the Research Board members and GoMRI awardees. As the consortia and small teams continued to improve the state of oil spill science and others showed interest in the conference, the Research Board realized that it needed to coordinate with other researchers who did not receive GoMRI funding but were deeply interested in the work it supported. The conference grew as the Research Board invited others to present their research at the conference. It would come to include representatives from industry; state and federal government officials; the media; nonprofits and NGOs; and spectators interested in the state of oil spill science. In its final years, the conference hosted about 850 participants and observers annually from diverse scientific fields and professional backgrounds. It had become a premier event for oil spill research. Colwell and the Research Board decided to rotate GoMOSES conference locations through the five Gulf states, providing more communities in the Gulf Coast region chances to attend the conference and learn about oil spill science.[33]

GoMOSES conferences built capacity in the region. They brought scientists and engineers from different disciplinary backgrounds and professional affiliations together to discuss science. Those in attendance shared results of their research. Rather than simply presenting solely at professional societies (such as oceanography conferences, ecology, and public health conferences), GoMRI provided a venue for science based on a much broader scale. Thus, attendees were provided with a comprehensive overview of the state of oil spill research and an opportunity to explore new directions in environmental and applied oil spill science.

Conclusion

Outreach, education, and capacity-building were three crucial components of the GoMRI model. Although the GoMRI mission was to fund the best science across its five themes, the Research Board understood that $500 million could make a significant difference in creating a legacy of scientific research and renewed public interest in the Gulf of Mexico. GoMRI's outreach, education, and capacity-building efforts broke down disciplinary boundaries and encouraged holistic approaches to oil spill science and for science in general. By offering venues for scientists to connect, GoMRI also overcame institutional differences, thereby building partnerships between schools and laboratories that could pool resources toward the common pursuit of the best possible relevant science, including those that focused on environmental quality issues such as oil spills. GoMRI also prepared hundreds of young scholars to build on their experiences working on an array of scientific issues in their postgraduate careers.

The Gulf region is likely to continue experiencing accidental spills because of the amount of oil drilling, refining, and transportation located there.[34] Deepwater Horizon engendered concerns among the local community that there were no scientific advocates willing to support them. GoMRI addressed this concern by increasing scientific capacity in the Gulf Coast and connecting nonscientific communities to the knowledge they sought. GoMRI funding also connected scientists and students from a range of disciplinary and institutional backgrounds. The GoMRI website will continue to be available through 2030 and the publications arising from GoMRI research, as Sea Grant publications will be maintained by Sea Grant and the Smithsonian portal. *Dispatches from the Gulf* will be available for the foreseeable future, on screen and on YouTube. GoMOSES continues in another form as a premier gathering for oil spill scientists and related professionals. Despite the Gulf region's propensity for oil spills and other disasters, GoMRI leaves behind an improved scientific capacity to respond to future such occurrences.

CHAPTER FIVE

GOMRI REFLECTS
Synthesis and Legacy

We realised early on that we would be producing thousands of datasets and research papers, and that somehow we needed to make the information that these contained more easily accessible, to people with widely varying interests. So we organised our researchers into collaborative teams covering eight core areas of interest, and charged them with producing documents that synthesised the information and made it comprehensible, to other researchers, to potential users, and to members of the public. We assigned a couple of Research Board members to each core area too, to act as "motivators" and help to keep the whole exercise on track. In the end we managed to condense all this into a single volume Special Issue that aims to bring it all together, and provide an accessible entry point for anyone interested into the wealth of literature that had been created. We may not have succeeded in this dream, but we gave it our best shot, and had some fun trying!

—DR. JOHN SHEPHERD, University of Southampton[1]

At the end of any major collaborative research program it is useful to take stock of what was learned. Such efforts are called synthesis in scientific parlance, and help reveal questions that should have been asked, gaps in scientific knowledge, and new methods to apply in future work. Synthesis also offers an opportunity to share the new scientific knowledge with nonscientific audiences. An article published in the journal *Environmental Science and Policy* defines *synthesis* as "the integration and assessment of knowledge and research findings pertinent to a particular issue with the aim of increasing the generality and applicability of, and access to, those findings."[2] In short, synthesis captures the current state of scientific knowledge and translates the knowledge to other communities.

Synthesis can be helpful as scientists work to translate the knowledge of their respective fields for application by user communities. Scientific publications are written for fellow practitioners. Nonscientific audiences may not be able to decipher how scientists use methods to analyze data and thereby reach conclusions. The scientific process— built as it is on argumentation, interpretation, and consensus among practitioners in the field—can be opaque to lay audiences. In some cases, researchers can have difficulty communicating scientific knowledge even to science-literate readers.[3] This becomes an obstacle when issues of global or regional significance require action from lawmakers or other stakeholder groups. GoMRI synthesis efforts were guided by the recognition that scientific knowledge should be made useful for those who are responsible for responding to disasters of major social, economic, and political significance.

According to the GoMRI synthesis and legacy webpage, "the overarching goal of the GoMRI's Synthesis effort is to document and exploit scientific achievements and advances, with the idea that synthesis will lead to new understanding and improved practices."[4] That overarching goal can be divided into three broad objectives. First, the Research Board intended that synthesis would survey and build on existing knowledge of oil spills in the Gulf of Mexico, a perennial problem for the region. Second, GoMRI used synthesis to translate the state of knowledge of oil spills in the Gulf of Mexico for nonscientists. Third, GoMRI's synthesis efforts helped inform better policy decisions pertaining to the role of oil in the Gulf. These efforts consisted of more than simply publishing a list of factoids learned through past research. Rather, GoMRI synthesis was meant to point a way forward, offering scientists new or under-addressed areas of inquiry to strengthen the overall field of oil spill science. To do so, GoMRI took stock not just of the knowledge that its funding generated, but the general state of oil spill science, regardless of which institutions were supporting research.

In addition to synthesis, the GoMRI Research Board worked for nearly five years on meeting its legacy goals, as stated by the Research Board in 2015:

1. Significantly advance the scientific understanding of the Gulf of Mexico, including its interactions with oil, dispersant, and dispersed oil by fostering the highest quality of research for the benefit of those who depend upon its ecosystems for their well-being.
2. Engender improved understanding, confidence, and trust of the public and other stakeholders (including the business and industrial community, government, policy makers and managers/planners, media, K–12 educators, undergraduate and graduate students as future scientists, GoMRI research participants, and the national and international scientific community) and inform best science-based policy and management.
3. Build intellectual capacity by: a) advancing relevant technologies; b) fostering research connectivity; c) building and maintaining an interactive GoMRI Gulf of Mexico database that can serve as a future baseline and to inform a more efficient future response; d) informing and training future scientists and engineers; and e) stimulating interest in STEM for K–12 students and educators.

4. Demonstrate that the responsiveness of the GoMRI model is appropriate and effective in serving the public good by enabling and overseeing timely and independent research funded through a private-public partnership with industry.

Because GoMRI had a built-in deadline marking the end of its official operations, the Research Board established clear, specific, and measurable benchmarks to achieve by the end of 2020.

Synthesis: More than a Summary

GoMRI synthesis began when the Research Board Synthesis and Legacy Committee decided to develop questions to be asked across all GoMRI scientific themes. They were:

1. What was the state of oil spill science that is, the baseline, prior to Deepwater Horizon?
2. What has the scientific community learned since Deepwater Horizon?
3. What major gaps in the knowledge still exist?
4. How can what we have learned be best applied?
5. Where do we go from here?

The GoMRI synthesis questions were intended to survey the state of oil spill research toward the end of GoMRI (questions 1–3), to translate that research for nonscientific audiences, and to inform policymakers how best to mitigate effects of oil spills in the Gulf of Mexico and in other regions (questions 4 and 5).

In addition to its guiding questions, the Synthesis and Legacy Committee developed eight core areas within which to categorize the knowledge that synthesis produced. The five questions would be applied to each of the eight core areas to determine baseline knowledge for the areas of research that GoMRI had provided support. The GoMRI website described the eight core areas as follows:

- GoMRI Synthesis Core Area 1, **Plume & Circulation Observations & Modeling**, focused on several physical oceanography areas of interest. Teams of GoMRI researchers collaborated to highlight research and modeling advances, and their relevance for oil transport and fate, in: 1) transport processes in the Gulf of Mexico along the river-wetland-estuary-coastal-open ocean continuum, 2) near-field, mid-field, and far-field dynamics of the plume, and 3) small-scale/near-surface/sub-mesoscale observations.
- GoMRI Synthesis Core Area 2, **Fate of Oil & Weathering: Biological & Physical-Chemical Degradation**, reviewed research advances related to several areas: methods of oil spill chemical analysis, use of genomics and proteomics, use of molecular biology tools, general exposure studies and techniques, photochemical reactions at sea and on shorelines, microbial degradation, Marine Oil Snow Sedimentation and Flocculent Accumulation (MOSSFA), and alternative dispersant research and technologies.
- GoMRI Synthesis Core Area 3, **Ecological/Ecosystem Impacts**, summarized research and identified gaps related to ecosystem impacts and the vulnerability and resilience of species and ecosystems to large-scale contamination events. Cross-cutting ecological impacts were investigated for several ecotypes: open-ocean, deep benthic, continental shelf, and coastal in-shore.
- GoMRI Synthesis Core Areas 4/5, **Human Health, Socioeconomic Impacts, and Ecosystem Services**, translated scientific findings to develop a conceptual framework for an Operational Community Health Observing System for the Gulf of Mexico states. The concept of allostatic load

was applied to the practical measurement of exposure to oil spills, further operationalizing new assessment techniques.
- GoMRI Synthesis Core Area 6, **Microbiology, Metagenomics & Bioinformatics**, generated foundational ideas to advance marine microbiology and enhance understanding of microbial response to major disturbances in the ocean.
- GoMRI Synthesis Core Area 7, **Integrated/Linked Modeling System**, focused on operational oil spill modeling, including ocean, wave, and weather forecasting, and development of a systems dynamic model, which could be used for future decision-making.
- GoMRI Synthesis Core Area 8, **Knowledge Exchange with User Communities: Lessons Learned and Operational Advice**, served as an advisory group that assisted in identifying lessons learned, promoted the effective application of GoMRI research results, and improved operational advice. The advisory group was comprised of representatives from academia, industry, the response community, government agencies, the NGO community, and other critical stakeholder groups. The members of this advisory group were routinely informed, and provided input to, the activities and results of other Core Areas.

The Synthesis and Legacy Committee developed the core areas in order to coordinate and streamline effectively efforts to answer the five synthesis questions. They differ from the GoMRI five research themes stated in the Master Research Agreement, though with key similarities. Core areas 1, 2, 3, and 4/5 matched closely with the five scientific themes described in chapter one. Core areas 4/5, 7, and 8 focused on how to translate knowledge about oil spills in the Gulf of Mexico to different user communities. They were key aspects of GoMRI's work to inform user communities about the role of oil in the Gulf of Mexico and its goal to improve future oil spill responses.[5] Core area 6 also deviated from the five research themes.

Recognizing there had been significant advances in the fields of microbiology, metagenomics, and bioinformatics, the Synthesis and Legacy Committee dedicated part of its synthesis efforts to those areas as well.

Core area 8 focused on translating knowledge to nonscientific audiences and ensuring GoMRI-funded research would be used to guide future oil spill responses. It brought scientists from industry, NGOs, and government together with nonscientific communities. As such, its efforts were organized and administered differently than the other core areas, which were overseen by scientists and researchers. Although it emerged at the same time as the other core areas, over the following years core area 8 came to function more as an advisory group that facilitated knowledge exchange among diverse stakeholders.[6]

GoMRI used core area 8 to advise the Research Board, identify what lessons had been learned, "promote the effective application of GoMRI funded science," and "improve operational advice."[7] Cochaired by Research Board members Dr. John Shepherd and Dr. Rick Shaw, the core area 8 advisory group members were drawn from a wide array of organizations and agencies representing some of the diverse stakeholders able to benefit from GoMRI's synthesis. The core area 8 advisory committee included representatives of oil spill response practitioners, NOAA, the petroleum industry, nongovernmental organizations, and academia, along with three stakeholders from the US Navy, the Canadian Department of Fisheries and Oceans, and Sea Grant.[8]

Core area 8 maintained a two-way conversation between the scientific community and user communities. Its advisory committee acted as both clearinghouse for scientific knowledge and translator of that knowledge for diverse groups. It had four main purposes: share updates on synthesis and legacy findings with colleagues; identify experts who could contribute to synthesis and legacy efforts; facilitate and contribute to effective communication between scientists and user communities; and advise on key issues in order to organize GoMRI synthesis workshops addressing various users' interests.[9]

Each core area incorporated multiple subtopics that fit under the umbrella of their main subject, with the exception of core area 8. For each subtopic, the core areas produced publications for specific scientific audiences in response to the five synthesis questions. In turn, researchers used the highly specialized publications to prepare a more general product for scientists in all fields.[10] Finally, scientists and researchers across all core areas used scientific publications to prepare products for the general public.[11] GoMRI synthesis involved creating highly specialized publications for specific scientific audiences, using those to produce interdisciplinary syntheses, and then using the interdisciplinary scientific literature to publish for nonscientific audiences. GoMRI synthesis thereby moved from the specific and particular to the more general and accessible. Core area 8 in turn "promote[d] the effective application and communication of GoMRI funded science between research and user communities."[12] In other words, it emphasized the fourth and fifth synthesis questions, ensuring the products reflected the interests of the user communities or alerted the user communities to issues in which they might be included.

GoMRI synthesis met the three objectives for successful synthesis as described in the beginning of this chapter. They synthesized new knowledge based on existing information, translated that knowledge for various stakeholders based on what those stakeholders identified as important issues, and created knowledge informing future policies related to oil drilling, refining, and transportation in the Gulf region.

Structuring the Synthesis

GoMRI synthesis and legacy changed over time but began early in the initiative's history. In October 2012, based on prior experiences, Colwell recognized the need for synthesis when GoMRI would

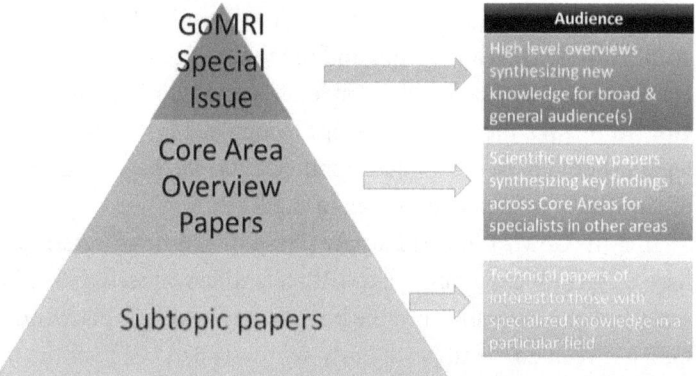

Figure 5.1. This diagram depicts the three-step process by which the core areas synthesized their work. Most of the products were oriented toward individuals with specialized knowledge in particular fields. Those products, however, were used to create higher level review papers in each core area for the broader scientific community. Finally, the intermediate Core Area Overview Papers were synthesized and used to write publications for general audiences. The main product at the highest level was a special issue of *Oceanography*, released in 2021. https://gulfresearchinitiative.org/gomri-synthesis/products/.

end its work. Two months later, in December 2012, members of the Research Board formed the Legacy Committee to begin developing a plan for describing GoMRI's enduring influence on oil spill science. Less than a year later, in October 2013, the board established the Data Synthesis Committee, chaired by Dr. Peter Brewer. Formation of both committees gave a firmer structure to synthesis and legacy efforts, but the Research Board quickly recognized the overlapping responsibilities of the two committees. In 2014, it decided to combine the two committees to create the Synthesis and Legacy Committee, responsible for organizing knowledge synthesis for the Gulf of Mexico oil spill and articulating the legacy of the initiative in 2020. Soon after its formation, the Research Board allocated $2 million to the Synthesis and Legacy Committee, which used the funds to support workshops and synthesis products. In addition, the Synthesis and Legacy Committee recommended that future RFPs include synthesis as a criterion on which they would be assessed.[13] Colwell and the Research Board understood that the public would be eager to know what GoMRI was able to learn and accomplish with $500 million and ten years of scientific research. Having begun discussion about synthesis and building a legacy for GoMRI in 2012, the Research Board recognized that synthesis and legacy should have been "incorporate[d] into the beginning fabric of GoMRI" and included in each of the RFPs.[14]

In 2017, GoMRI entered a new phase of synthesis, referred to here as "scientific synthesis." GoMRI's scientific synthesis differed in scale and organization from the board's data synthesis discussions, held between 2011 and 2016. Early synthesis had been conducted through periodic workshops. After 2017, GoMRI undertook synthesis through both top-down and bottom-up approaches, including hundreds of scientists and members of various user committees. The later synthesis effort had a more comprehensive framework and asked a definitive set of questions. As the RFP process ended in 2017, the Research Board began to dedicate much of its remaining time and funding to assessing what had been learned and how that knowledge could be applied in the field. Soon after, the GoMRI five synthesis questions and eight core areas were established.[15]

The Research Board was then confronted with the question of who would be best equipped to undertake the final work of synthesis. Researchers themselves knew the most about the state of their fields. However, the GoMRI Research Board had taken a bird's-eye view of progress within the field of oil spill science since 2010. The Research Board organized synthesis in two parallel units, with the GoMRI's Research Board responsible for initiating the top-down and consortia and the individual investigator teams organizing their own bottom-up synthesis, as stipulated in RFP-VI. Each core area had an advisory committee to organize workshops and review synthesis products. This parallel model worked successfully and GoMRI was able to take an approach to synthesis that produced both highly specific and general publications for the wide range of its audience.

As noted above, the Research Board played an important role in synthesis. Because the objective of GoMRI synthesis was to survey and build on the best knowledge available, synthesis was not centralized under the Research Board's control. Rather, Research Board members were brought into synthesis as motivators, helping to move along synthesis in the fields in which they had expertise. The motivators coordinated and communicated with synthesis advisory committees to ensure the syntheses were not duplicated or redundant.[16] The Research Board was important in ensuring synthesis focused on the five synthesis questions. The Research Board essentially served as coach and umpire, motivating advisory committees to meet their obligations and ensuring all core area work fell within the framework that the Synthesis and Legacy Committee developed.

To organize the synthesis and the community to survey, build on, and translate the state of scientific knowledge about oil spills, the Research Board supported a series of workshops. Workshops had been the main vehicle for its synthesis publications since 2012, but in subsequent synthesis, additional workshops came to cover a range of subfields within the core areas. Initially envisioned as a series of a half dozen workshops, the number grew as the GoMRI synthesis evolved. To keep the workshops manageable and allow communication and coordination among the participants, the workshops were limited to about thirty people. In keeping with its focus on building capacity and supporting early-career scientists, the GoMRI Synthesis and Legacy Committee strongly encouraged the participation of recent graduates and mid-career scientists in the synthesis workshops.[17]

To motivate—but not control—the synthesis, the GoMRI Research Board issued guidelines for building successful workshops. Each synthesis workshop would be organized around three or four key topics or questions related to one or more of the GoMRI research themes and developed by a lead team of six to eight people, including a Research Board member and one or two members of the user communities. Core area 8 ensured expert scientists and stakeholders benefitting from new oil spill research were included in organizing the workshops. Notably, stakeholders other than scientists (representatives from academia, government agencies, industry, and NGOs) were invited to join the organizing committees. Before each workshop the organizing committees circulated relevant publications and listed key topics and questions for attendees to discuss.

The workshop agenda listed organizers presenting introductory comments citing goal setting and plenary sessions and orientation for participants. These included a summary of current knowledge and findings. Plenary sessions covering key topics and questions were followed by breakout sessions to discuss questions or topics that had been covered. On the third day, workshop rapporteurs were assigned to present findings from breakout sessions after which reports of the workshop were drafted.

Research Board guidelines for each synthesis workshop included criteria for the synthesis product, essentially the initiative's five questions:

1. The known (Status of core area prior to Deepwater Horizon)
2. From unknown to known (Evolution of core area since Deepwater Horizon)
3. The currently unknown (Where science needs to go from here)
4. What has been learned (How we can best apply new knowledge)
5. What should follow next? (And who and how?)
6. Conclusions and recommendations (Major findings and gaps that needed to be addressed)

The Research Board provided a guide to organizing synthesis workshops and recommendations on structuring the report. As described above, however, synthesis was community driven, with each workshop involving participation of many research scientists from both within and outside the GoMRI community.

What Was Learned?

The primary purpose of this manuscript is to provide a history of how GoMRI came together and developed over the course of its decade-long operation, so that the lessons learned from its operation can be applied to other scenarios where science must be conducted during a disaster. Still, this manuscript would be incomplete without at least a partial accounting of what, exactly, the $500 million that GoMRI provided to researchers accomplished. In 2021, *Oceanography* magazine published a special issue titled *Ten Years of Oil Spill Science*. It was a major synthesis project, with fifteen articles detailing the many discoveries, innovations, and findings produced with GoMRI research funding. This section draws from that special issue to provide a brief overview of GoMRI's technological, scientific, public health, and socioeconomic accomplishments. Readers interested in a fuller accounting of what GoMRI accomplished would be well advised to direct their attention to the *Oceanography* special issue.[18] They might also turn their attention to the different synthesis products that each core area generated, which can be found in the citation at the end of this sentence.[19]

Technology

GoMRI funding allowed researchers to employ a wide range of technologies to examine the effects of the Deepwater Horizon oil spill. Satellites proved to be especially useful in studying the ways that oil moved across the surface of the Gulf of Mexico. The surface slick that the blowout produced covered an area of about 11,200 square kilometers—larger than the state of Delaware—for three months, with the total footprint of the slick changing hourly as the wind and undersea currents carried the oil in different directions.[20] Traditionally, satellite and aerial photography have helped responders figure out where oil slicks move. Photographs only capture changes to an oil slick taking place on the surface of a body of water, however; their target can also be obscured by cloud formation. GoMRI funding helped improve oil slick observations. GoMRI's ECOGIG consortium, along with the NOAA National Resource Damage Assessment (NRDA) program, together supported the use of Synthetic Aperture Radar imaging to study the dynamics of oil slicks in the Gulf of Mexico. Unlike optical imaging, SAR can penetrate clouds and allow scientists to obtain data about other characteristics of the Earth's surface, such as moisture and structure.[21] Using SAR also helped scientists understand in greater detail other aspects of the oil slick, including its thickness in different locations and the degree of emulsification.[22] One of the characteristics of an oil spill that makes it suitable for SAR imaging is that oil creates a smoother surface that deflects radar energy more effectively than unoiled surface water.[23]

During previous major spills, scientists would have to wait several days to see a processed satellite image and understand how the oil was moving across the sea's surface. SAR can be processed in a matter of minutes, granting responders close to a real-time understanding of how the slick was behaving in the Gulf of Mexico.[24] One crucial finding was related to the use of chemical dispersants in response to Deepwater Horizon. Scientists discovered that while "the submarine dispersant application may have achieved a reduction in the total volume of surface oil," it also increased "the area over which the residual volume was dispersed."[25] As Michel Boufadel and others put it in their article about SAR, "it is hoped that this experiment in satellite remote sensing of an oil spill will not be repeated, but the data obtained were and will remain an important resource for understanding the dynamics of floating oil in the marine environment."[26]

SAR allowed scientists and first responders to better understand the dynamics of oil spills at or near the surface of the Gulf. Other new technologies were deployed to gather data about how oil moves throughout different sea levels, including just below the surface and closer to the sea floor. The Consortium for Advanced Research

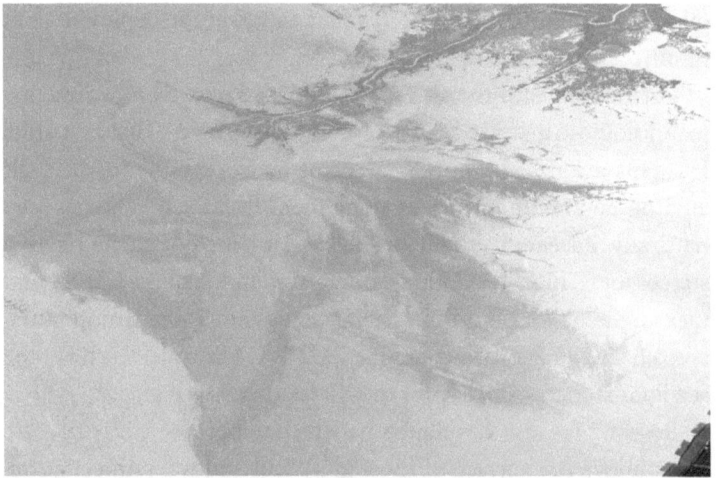

Figure 5.2. A JAXA astronaut took this picture of the Deepwater Horizon oil spill while aboard the International Space Station. This image was captured on May 4, 2010, about two weeks after the blowout. Photo by Soichi Noguchi, courtesy of NASA.

on Transport of Hydrocarbon in the Environment (CARTHE) deployed several devices to track the movement of oil throughout the Gulf environment. In 2012, CARTHE conducted the Grand Lagrangian Deployment (GLAD) to study the role of submesoscale currents (those that take place at the medium level between the surface and the deep sea) in transporting oil through the Gulf of Mexico.[27] GLAD relied on an easy-to-assemble CARTHE drifter, which consisted of a floating disc containing a GPS locator powered by batteries connected to a submerged drogue that would allow the drifter to remain upright. The drogue was pulled by below-surface currents. The unit was made from a biodegradable material that reduced ocean pollution.[28] Three hundred drifters were released over a span of ten days during the first GLAD. The drifters could update their location every five minutes for up to six months.[29] GPS data from the drifters showed that small vortices and fronts tended to lead them to cluster in particular locations, suggesting where large portions of the oil might travel in the event of another major Gulf oil spill. Responders can use this data to target the deployment of resources during the next such disaster.[30]

Further beneath the surface, scientists deployed new imaging technologies to study proxies for the Deepwater Horizon spill. Humans are not the only sources of hydrocarbons in the Gulf environment. In many places along the Gulf floor, oil and gas seep naturally. Researchers at Texas A&M University attached a new stereoscopic, high-resolution camera system to collect images and measurements of oil droplets and bubbles emerging from natural seeps.[31] The images and data that TAMU-CAM collected from the sea floor stood as a proxy for the Macondo oil well seep.[32]

Researchers also developed new technologies to study the fate of oil above the surface of the Gulf. Wind and waves meet at the surface of the Gulf of Mexico. As an oil slick grows, oil droplets can become aerosolized through wave action, then carried into the atmosphere. Under certain conditions, aerosolized oil can be lifted high into the air, threatening birds and other wildlife.[33] To study this phenomenon, scientists at Johns Hopkins University Laboratory for Experimental Fluid Dynamics built custom wave tanks capable of simulating a variety of surface conditions. GoMRI also supported a related effort to study the effects of aerosolized oil on human lungs. Another research team at Johns Hopkins introduced a new type of exposure chamber to study how aerosolized oil droplets affected lung tissue cultures.[34] Their initial results revealed that there were "similar initial trends in cellular changes from cigarette smoke."[35] Their research efforts supported a health risk assessment for workers who were on the water responding to the spreading oil slicks, and will prove useful in future response efforts.[36]

Science

GoMRI funding greatly improved the state of scientific knowledge about oil in the Gulf of Mexico. Referring to the use of satellites to gather data on the expansion and flow of the oil slick, scientist Michel Boufadel and others wrote that "it is hoped that this experiment in satellite remote sensing of an oil spill will not be repeated, but the data obtained were and will remain an important resource for understanding the dynamics of floating oil in the marine environment."[37] Deepwater Horizon was an undeniable disaster, but the ability to quickly harness institutions, individuals, and technologies to study the spill will allow the region respond to the next major oil spill disaster.

Satellite imaging and the CARTHE drifters both revealed how oil moved and spread through the upper layers of the Gulf of Mexico. Using GoMRI funding, scientists also learned how marine oil snow and flocculent accumulation play important roles in distributing oil and dispersants to the sea floor. Marine snow refers to particles greater than 0.5 millimeters in size consisting of "aggregations of smaller organic and inorganic particles, including bacteria, phytoplankton, microzooplankton, zooplankton fecal pellets" and other organic and inorganic materials.[38] Marine oil snow is an aggregate of marine snow and oil. When the substance flocculates—clumps together and forms larger clusters—scientists refer to it as Marine Oil Snow Sedimentation and Flocculent Accumulation (MOSSFA). As MOSSFA descends to the sea floor, it is "grazed by zooplankton, degraded by microbes, or finally integrated into sediments."[39] C-IMAGE, along with other GoMRI consortia, established a working group dedicated to the study of MOSSFA and its role in removing oil from the Gulf ecosystem. Their efforts revealed that approximately 14 percent of the 111 million liters of oil released during the Deepwater Horizon spill was distributed through the MOSSFA process.[40] As Farrington and others have written, GoMRI-funded research has added consideration of MOS and MOSSFA to oil spill response, modeling, and research.[41] In general, such research has improved scientists' understanding of where, exactly, the oil released from a major spill goes.

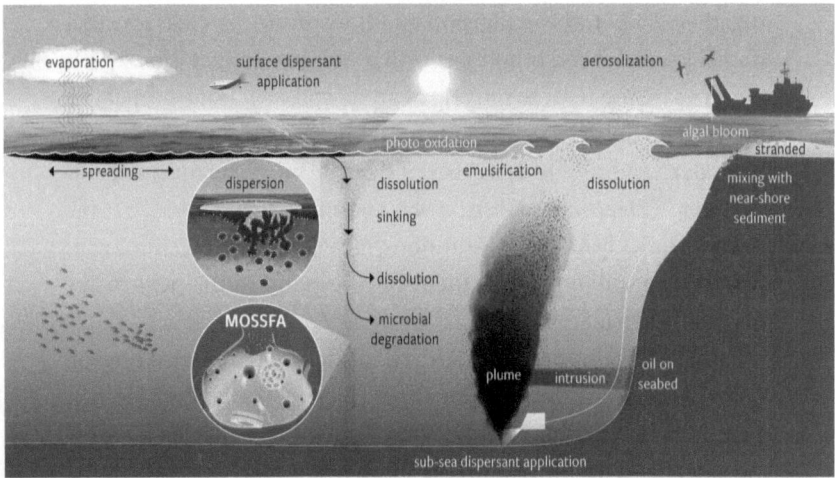

Figure 5.3. A figure that shows the many fates of oil spilled in a marine environment. Available in Antonietta Quigg et al., "A Decade of GoMRI Dispersant Science: Lessons Learned and Recommendations for the Future," *Oceanography* 34, no. 1 (March 2021): 98–111. https://doi.org/10.5670/oceanog.2021.119. Licensed under CC BY 4.0, https://creativecommons.org/licenses/by/4.0/.

MOS and MOSSFA are not inert substances when they reach the sea floor. Rather, they can interact with the surrounding ecosystem, sometimes bringing deleterious consequences. As Ken Halanych and others have pointed out, MOSSFA can harm coral reefs and the ecosystems they support. Coral colonies that were more heavily covered with MOSSFA were often severely injured and began a long period of decline as indicated by the loss, in some cases, of entire branches. Reefs that were lightly covered, on the other hand, have shown signs of recovery.[42]

GoMRI-supported research also made a strong case for photo-oxidation as a significant process in the breakdown and ultimate fate of spilled oil. Research conducted in the 1970s and 1980s revealed that petroleum was sensitive to the process of photo-oxidation, but as Chuck Wilson and other have noted, "the related effects were downplayed and grossly underestimated in terms of mass balance significance over more than a decade."[43] Researchers affiliated with GoMRI hope that the better understood significance of photo-oxidation in the breakdown of oil particles might influence future response tools to major oil spills.

Apart from influencing new response methods to major oil spills, GoMRI research also brought into question some previous methods of protecting shoreline ecosystems from oil spill damage. In particular, Deepwater Horizon revealed the immense trade-offs involved in the use of freshwater diversions to keep oil away from shorelines. The Louisiana Delta region is a place where freshwater carried from the Mississippi meets the saltwater of the Gulf of Mexico. It is a very wet region, though various agencies—most significantly, the Army Corps of Engineers—have tried to separate dry land from wet marshes, bayous, and swamps. Such efforts have resulted in a heavily engineered and managed landscape.[44] It is possible to divert additional freshwater into the Mississippi River and other spillways, thereby increasing the outflow of those sources. The idea behind freshwater diversion is simple: additional water diverted toward the Gulf of Mexico would help push oil away from coastal wetlands. Initial simulations by the Army Corps in late April 2010 demonstrated the feasibility of this approach, and flow rates from the Davis Pond diversion, which flows into Barataria Bay, were increased significantly after Deepwater Horizon. This was not effective, as Barataria and other parts of coastal Louisiana quickly "reached over three orders of magnitude greater than background contamination in some locations and remained several orders of magnitude higher than baselines for at least five years thereafter."[45] Species in the boundary zone between fresh and salt water environments have evolved to withstand a certain range of changes in salinity, but the sudden influx of fresh water due to the Army Corps diversions greatly increased oyster mortality—even more than the oiling that they were meant to prevent. Many more ecological changes will likely take years to fully understand.[46]

Dispersants

The initial response to the Macondo well blowout involved the use of chemical dispersants injected directly at the broken wellhead, as detailed in chapter one. This practice was unprecedented, both because of the delivery method that responders used and because of the volume of dispersants that were both sprayed on the Gulf surface and injected at the wellhead. Approximately 4.1 million liters of chemical dispersants were sprayed on the sea surface; another 2.9 million liters were injected at the wellhead.[47]

GoMRI funding produced 176 publications that focused on the use of dispersants in response to the spill.[48] Much of the research opened areas of inquiry that future scientists may wish to investigate. When dispersants were applied during Deepwater Horizon, the complexity of the Gulf environment and ecosystem proved difficulty to re-create in a laboratory setting. Most publications focusing on laboratory studies of dispersants—128 as of 2021—rather than field measurements and observations (18), left open many questions about chemical exposure rates, the potential hazards of combined oil and dispersant mixtures, their effects on marine life and human health, and how dispersants influence the formation of MOS and therefore the ultimate fate of oil released during a spill.

In their synthesis article for *Oceanography*, Antonietta Quigg and her coauthors identified three grand challenges that GoMRI research on dispersants brought to the foreground. The first is on the use of subsurface dispersants injected during a major spill. GoMRI supported the development of experimental setups that could simulate conditions during the Deepwater Horizon spill to determine how subsurface dispersant injections affected the formation of oil droplets, which could shed light on the ultimate fate of oil and dispersant mixtures and how quickly the oil might biodegrade.[49] The second grand challenge applies to experimental design. GoMRI research has drawn attention to the different ways that experiments must be designed and carried out in order to reach any firm conclusions about issues such as dispersant toxicity. The third challenge is to research new dispersants. In response to the use of dispersants during past spills, such as *Exxon Valdez*, companies introduced new dispersants that promised to be effective without the incorporation of the most hazardous chemicals. GoMRI research demonstrates that research on the potential benefits and harms of newly introduced dispersants needs to continue.

GoMRI research brought more attention and focus to the question of how best to utilize dispersants and how to improve their use in response to future disasters. Still, many questions remain about their efficacy, methods of distribution, and potential hazards. Quigg and her coauthors put it best when they wrote:

> It will . . . take time and research to determine whether the dispersants themselves, used in such high volumes and at subsea, are in fact effective at what they are intended to do and whether they have any longer-term detrimental effects on marine life and/or public health. Along with the grand challenges identified above, there are further topics that require additional attention. Last, but not least, during GoMRI, as noted above, research has been undertaken to explore alternatives to the established dispersants . . . in the United States. As more knowledge is gained about alternative formulations, and perhaps a much more effective and less harmful dispersant is discovered/formulated, there is the pragmatic issue of Corexit 9500A and other dispersants currently listed on the National Contingency Plan Product Schedule that are already purchased and stockpiled in key locations. . . . There remains a paucity of information on the long-term consequences of dispersants in the marine environment, as little is known about the fate of household cleaners and products such as shampoos and dishwashing liquids. Thus, the use of these dispersants enters the realm of the interfaces of science-economics-policy management. We submit that the

current existence of a stockpile of prepositioned dispersants should not hinder research that could lead to more efficient and potentially less harmful dispersant formulations.[50]

Microbial Genetics

GoMRI research greatly improved the state of knowledge about microbes, microbial communities, and oil spill environments. Microbes can play a significant role in degrading the hydrocarbons released during oil spills. "Despite their importance," Shannon Weiman and her coauthors wrote in *Oceanography*, "relatively little was known about marine microbes that degrade hydrocarbons in the Gulf of Mexico prior to the Deepwater Horizon spill, nor the effect of hydrocarbons on the microbiology of the Gulf region."[51] As was the case with the public health research described above, GoMRI played an important role in filling a significant gap in scientific knowledge about the role of microbes in oil spill environments. This was reflected, quite basically, in improved understanding of the types of microbes that tend to thrive in hydrocarbon-rich environments.

An important discovery that GoMRI generated was that microbial communities could work together. As Weiman and her coauthors write, "GoMRI research led to the discovery of metabolic partnerships for oil degradation . . . [revealing] that genes within a single metabolic pathway may be distributed throughout the community rather than contained within a single microbial species or strain."[52] These discoveries might contribute to future oil spill responses. Wieman writes that "by examining the presence of oil-degrading species, genes, and biogeochemical pathways, scientists can assess the natural bioremediation potential of a community, providing insight into whether and how its metabolic capabilities can be augmented."[53] Some recipients of GoMRI funding have proposed using nitrogen fertilizers to accelerate the microbial degradation of oil. GoMRI has also generated promising research about the potential use of microbes, rather than chemical dispersants, as a nontoxic and biodegradable means to dissolve oil in the Gulf.[54] Wieman and her coauthors were hopeful about this discovery, writing "strategies to promote growth of native microbial species or utilization of genetically engineered microbes for natural surfactant production would provide responders with biodegradable, nontoxic surfactant alternatives. Natural microbial surfactants may also prove effective as nontoxic dispersants with fewer environmental concerns."[55]

Public Health

Prior to GoMRI, little data and knowledge were available at all about the public health and socioeconomic effects of major oil spills. The baseline knowledge was simply not there. There was some information about how previous spills had affected US Coast Guard personnel, but as Chuck Wilson and others wrote in the special issue of *Oceanography*, "researchers were faced with the reality that baseline health and socioeconomic data against which to compare spill effects were woefully inadequate. As a result, this scarcity of systematically collected and curated health and socioeconomic data made it difficult to characterize individual, community, and societal impacts and to connect outcomes with their causes."[56] Much of what was new about GoMRI's public health and socioeconomic research, then, was simply improving baseline knowledge about how the oil spill affected the communities closest to and most reliant on the health of the Gulf to sustain their livelihoods. GoMRI was able to significantly improve the baseline understanding about the relationship between public health and major oil spills by providing continuous funding to monitor and collect data about local communities and the consequences they faced from the Deepwater Horizon oil spill.

As a result of GoMRI's support for public health and socioeconomic research, public health experts have strongly advocated for a Gulf of Mexico Community Health Observing System focusing on the sixty-eight coastal counties surrounding the Gulf. The effects of a single disaster on public health in a region both compounds the effects of previous disasters and are in turn compounded by future disasters. As Paul Sandifer and others point out, "environmental disasters of various kinds and magnitudes occur frequently in the Gulf region." Apart from Deepwater Horizon, the Gulf Coast continues to rebuild from Hurricane Katrina (2005), Hurricane Harvey (2017) and other disasters. Sandifer and his coauthors wrote that "recurrent disasters take a heavy toll on human health in the region, where many people already suffer significant health and economic disparities."[57] There is a vital need in such a disaster-prone region for systems that can track health and socioeconomic outcomes over the long term. As Sandifer and his coauthors write,

> Human health findings were severely limited by a lack of baseline health data, long delays in implementing major health research activities following the spill, heavy reliance on self-reported and cross-sectional survey data, limited collection of clinical health information, and a paucity of long-term, longitudinal cohort studies. Health studies need to be initiated before, during, or immediately following a large spill and must continue long enough to identify long-term effects and secondary surges of chronic illnesses. Critically needed are cohort studies that start before a major spill and continue through it and onward for a long period after.[58]

GoMRI demonstrated the utility of such efforts. As in the case of river diversion and its consequences for Gulf ecosystems, the effects of Deepwater Horizon and other disasters on the region will be borne out in the long term. Establishing a robust monitoring system will help demonstrate which communities are most vulnerable, and help responders figure out how to help them recover more quickly.

Another major finding of GoMRI-sponsored research on public health and socioeconomic effects was that there is a dire need for scientists and first responders to communicate their findings in clear and coherent language. According to Sandifer and his coauthors, "[w]hen fisheries were reopened following closures after the spill, Gulf seafood was demonstrated to be safe for human consumption within guidelines existing at the time. However, there was much uncertainty and persistent worry about seafood safety, even years after the spill. For future large spills that affect fishing zones, thorough and rapid appraisals of seafood safety should be undertaken immediately after the spills, followed by plain language communication regarding consumption risks based on appropriate demographic information."[59] GoMRI took its mission to educate and inform the public seriously. The research it funded on public health and socioeconomics confirmed the value of engaging the public throughout the scientific response to a major disaster.

Legacy

The Master Research Agreement stipulated GoMRI would terminate at the end of 2020. The Research Board had achieved consensus on four legacy goals to accomplish by the end of its operations. Those goals reflect the Research Board agreement to pursue the best possible relevant science and when generated it should be applied to environmental, economic, technological, sociological, and human health issues of the Gulf region. The goals were:

1. Significantly advance scientific understanding of the Gulf of Mexico, including interactions with oil, dispersant, and dispersed oil by fostering the highest quality research for the benefit of those who depend upon its ecosystems for their well-being.
2. Engender improved understanding, confidence, and trust of the public and stakeholders (business and industrial community, government, policy makers and managers/planners,

media, K–12 educators, undergraduate and graduate students as future scientists, GoMRI research participants, and the national and international scientific community) and inform science-based policy and management.
3. Build intellectual capacity by advancing relevant technologies, fostering research connectivity, building and maintaining an interactive GoMRI Gulf of Mexico database as a baseline to inform a more efficient future response, informing and training future scientists and engineers, and stimulating interest of K–12 students and educators in STEM.
4. Demonstrate the GoMRI model serves as an appropriate and effective source of the public good, enabling and overseeing timely and independent research funded through a private-public partnership of industry, academia, and the public.

The GoMRI legacy goals are significant for two reasons. First, they showed that GoMRI's paramount responsibility was to support the best possible scientific research. Second, the legacy goals revealed that GoMRI was interested in using that science to help the Gulf Coast rebuild from the Deepwater Horizon disaster and inform public policy to minimize to the extent practicable future such occurrences. GoMRI's legacy goals went beyond scientific research alone. They addressed the problem of oil spills and helped mitigate future disasters by connecting the public to scientific knowledge about the Gulf.

The first legacy goal described GoMRI's effort to understand the consequences of the largest marine spill in the history of the United States. The scale of the spill was a factor with the US Environmental Protection Agency approving use of thousands of gallons of dispersants, including dispersants deployed at the wellhead, an unorthodox response. While helping reduce slicks on the Gulf surface water and breaking down oil into smaller particles, unconventional use of dispersants to deal with the Deepwater Horizon spill necessitated research on both short- and long-term effects. Oil spill scientists understood relatively little about the effects of dispersants on the environment and ecology of the Gulf of Mexico. Thus, significant research on the role of dispersants was needed, not only to address the current need but also to inform future oil spill responses.

The second legacy goal showed that GoMRI was demonstrating the value of networks of scientists and researchers in building connections among diverse regional stakeholders. This would assist in any future oil spill responses, with first responders, policymakers, local industries, and the scientific community equipped with improved knowledge about the consequences of the Deepwater Horizon spill. An added benefit was the outreach effort to encourage interest in the Gulf of Mexico among the next generation of students. Partnerships between consortia and regional educators helped develop new curricula providing K–12 students with a much better understanding of the Gulf of Mexico and the environmental legacy of the Deepwater Horizon spill. As students progressed in their careers, they would benefit from the scientific networks and multi-institutional partnerships GoMRI had fostered and funded. GoMRI thereby had funded scientific research and also connected many audiences with the knowledge generated.

The third legacy goal addressed GoMRI's efforts to build capacity, discussed in chapter four. GoMRI's consortium funding model encouraged interdisciplinary and intermural networks of scientists collaborating on initiative themes. Indeed, one of the goals GoMRI achieved was "the improvements of knowledge, systems, techniques and practices that will continue to exist and be of value after the program has concluded in 2020."[60]

Finally, the fourth legacy addressed the expectation that the scientific knowledge generated would be used in the public interest. GoMRI established a robust review process for allocating the $500 million in funding. Based on the experience of the Research Board with NSF and other federal and state agencies and institutions, GoMRI crafted a strict code of conduct and culture of ethical science. Although the funding was not public but private, the process the

Research Board established ensured the best possible relevant science would be funded, notably without political or corporate influence.

Conclusion

GoMRI synthesis and legacy coalesced in 2012 as items of discussion at the first GoMRI management team meeting. Although initially separate, they were combined in 2014 with the merger of the data synthesis and legacy committees. In 2016, RFP-VI, the final call for proposals, included the requirement for synthesis. In early 2017, the newly formed Synthesis and Legacy Committee developed a framework to guide later synthesis efforts. When GoMRI operations were ending, the question of what had been learned from the investment of $500 million was raised by members of the Research Board as a grand challenge to meet by 2020. GoMRI synthesis became the mechanism for meeting the challenge. As GoMRI operations began to wind down, the core area leaders, Research Board, and management team prepared a comprehensive summary of the decade of research: the dedicated volume of *Oceanography* magazine, detailing the history of GoMRI and released in 2021.

In general, synthesis entails surveying and building on existing scientific knowledge, translating it for relevant communities and user groups, and informing public policy. GoMRI accomplished synthesis by bringing together scientists from within and outside the initiative funding structure. Members of the Research Board, scientists, and user communities collaborated in organizing workshops addressing concerns and questions of all participants as clearly and fully as possible. In turn, the workshops, initially envisioned as a few working sessions, evolved into a significant effort and produced publications reaching audiences across the country and around the world. GoMRI synthesis extended well beyond Deepwater Horizon.

The GoMRI legacy now includes an improved capacity to respond to oil spills in the Gulf region and beyond. In summary, the legacy improves trust between the general public and research scientists, with respect to oil spills. As long as oil remains a key energy source, spills will continue to happen, despite best efforts to reduce accidental inputs by way of regulations and increased attention to safety. By educating the public and proving a commitment to open and transparent research, GoMRI scientists leave a legacy of improved knowledge about the role of oil in the marine environment and greater public confidence in results of GoMRI research.

CONCLUSION

THE GoMRI MODEL
Social Considerations and the Best Science

GoMRI provides a model for how to build a large-scale scientific research endeavor, beginning with an appointed Research Board and resulting in a complex program, in a very short time. The Deepwater Horizon disaster afforded a rare opportunity to examine a particularly devastating oil spill. The scientific community wasted little time coming together to determine how best to dedicate funding, resources, and expertise to study the spill and its effects on citizens and their environment. With an overarching directive to fund the best possible relevant science, the GoMRI Research Board refined the five theme areas dictated by the Master Research Agreement. BP provided a total of $500 million to fund scientific research across the designated areas. Although not distributed across the themes precisely equally, the end product was a well-researched and balanced volume of scientific knowledge for each theme.[1] While supporting scientific research, GoMRI also created broad and far-reaching programs that will outlast the initiative and benefit the region for the foreseeable future.

Funding the Best Science

What was GoMRI's overarching goal? With unanimity, the Research Board would answer "to fund the best possible relevant science." Because GoMRI did not commit a set amount of funding to each of its scientific themes, the proposal review process was not focused on allocating an equal distribution of research funds to each theme. The Research Board chose to fund proposals based on their merit. Theme area three (environmental effects of the petroleum/dispersant system on the sea floor, water column, coastal waters, beach sediments, wetlands, marshes, and organisms; and the science of ecosystem recovery) received most of the funding. Theme area five (impact of oil spills on public health including behavioral, socioeconomic, environmental risk assessment, community capacity, and other population health considerations and issues) received the least because of the paucity of proposals submitted in response to the RFP. Although the Research Board did its best to balance the "bouquet of funding" across the five themes, ultimately its commitment to supporting the best science resulted in some themes receiving more funding than others.

To ensure the process for allocating funding was fair, ethical, and based on merit, the Research Board partnered with the Consortium for Ocean Leadership in 2011. Ocean Leadership had experience in drafting requests for proposals and reviewing the proposals received in response to an RFP. GoMRI reviewed the language and evaluation criteria for the small team and consortium RFPs, which were issued between 2011 and 2016. As discussed in earlier chapters, the Research Board was open to revising and rewording the RFPs, based on what had been learned from earlier proposal solicitations. The process of funding the best science was not, therefore, predetermined, but was, in fact, a product of each year testing and observing what worked and what improvements were needed, while adhering to an overall framework for review and ranking. As the Research Board identified and addressed issues and ideas in its policy memoranda and revised RFPs,

GoMRI was able to ensure the scientific research that was funded would undergo improvement with each round of funding solicitation.

Science Advocacy: A Resource for Residents of a Disaster-Prone Region

Oil spills after Deepwater Horizon remain a constant threat to residents and industries located in the Gulf Coast region. Offshore platforms and onshore oil tanks spill approximately 330,000 gallons of crude oil each year offshore of Louisiana alone. The complex interaction of environmental forces and oil infrastructure has caused dramatic and destructive disasters throughout the history of the Gulf oil industry. An oil spill sometimes happens very suddenly, as was the case with Deepwater Horizon, or very slowly over time, as in the case of the Taylor Energy oil spill, which has leaked into the Gulf continuously since 2004.[2] Since 2004, there has been a total of five major oil spills in the northern Gulf of Mexico and along its shoreline, with over a half million gallons released and countless smaller spills.[3]

In the immediate aftermath of Deepwater Horizon, BP hired several scientific teams and chartered dozens of vessels to conduct field research. The company sought to determine the quantity of oil flowing from the damaged wellhead. Independent scientists studying the spill reported that BP's analysis of the volume of oil released per day was far less than the actual amount. While BP reported a spill rate of 1,000 gallons per day, soon after the independent scientists reported a rate closer to 5,000 gallons per day. In mid-May, National Public Radio reported that the spill rate was an order of magnitude more in the first weeks than what had been calculated by BP. By then, the independent scientists had reached a consensus that the spill rate was closer to 50,000 gallons per day.[4] The misestimation by BP of the volume of oil escaping daily from the broken wellhead did little to engender trust between the local community and the company. In 2015, a shrimper, Dean Blanchard, saw his business decline by 85 percent during the five years since the spill and stated his frustration and anger succinctly. "I mean who are you going to call over here? You going to call your senator? Or your representative? They've been on BP's payroll since they've been in politics."[5] Blanchard's anger and resentment toward the company and politicians perceived as catering to the oil industry spoke for many in the Gulf region.

Blanchard's comments reflected a deeper sense of vulnerability on the part of Gulf Coast residents. In the wake of Deepwater Horizon, as with many spills before it, communities around the Gulf region sought answers about the state of their fisheries, the quality of local bodies of water, and other pressing environmental matters from regional scientists, state and federal government agencies, and NGOs. For outreach and education, GoMRI gathered regional educators and students into teams often quite directly by providing teachers an opportunity to participate in field research. The GoMOSES conferences were open to residents of all five Gulf states to attend the scientific meetings and learn about the research being conducted on the Gulf of Mexico. GoMRI and its partner organizations stepped up to answer many of those questions and help the community understand the challenges it faced.

Sea Grant proved to be a key partner in these efforts to connect with regional communities. Sea Grant was able to closely monitor the research being done on the science of oil spills in the wake of Deepwater Horizon. It was able to seek answers from GoMRI-funded researchers and carry the information back to local communities and user groups. By conducting the most recently developed and effective methods of surveying regional communities, such as social network analyses, as well as the more traditional workshops and panels, Sea Grant came to learn, with some precision, what Gulf Coast communities needed to know about the state of the Gulf of Mexico. Because Sea Grant had the means to translate scientific findings to lay audiences, many Gulf residents and communities came to feel as though scientists were working on their behalf. Sea Grant was effective in translating the range of knowledge that GoMRI produced back to local communities to respond to their questions about the consequences of the Deepwater Horizon disaster. It was a vital bridge spanning lay audiences and the GoMRI community of scientific researchers.

Oil spills like Deepwater Horizon have shaped the culture of the Gulf Coast over the course of the twentieth century. As historian Andy Horowitz writes in his environmental history of the Gulf region, "for people on the coast, what could appear on the surface to be an acute, chemical event, became absorbed into a chronic and diffuse cultural trauma."[6] Horowitz was writing about the Deepwater Horizon disaster, stating that long after the slicks were gone from the surface of the water and the media had moved on to other matters, the incident would linger in the culture, economy, and memory of the residents of the Gulf Coast region. Indeed, this is true of many such accidents in the Gulf region. GoMRI helped to advocate for the interests and needs of Gulf residents while working to understand, rehabilitate, and protect the shorelines that are central to their lives. By continuing to research the effects of the Deepwater Horizon spill for a decade after it happened, GoMRI ensured that the scientific community treated it as a chronic and diffuse event, rather than an acute and short-lived crisis.

The sense of scientific advocacy that the GoMRI Research Board carried into its work was a part of what it meant to fund the best possible relevant science. Of course, scientific knowledge can be produced without taking into account the needs and concerns of communities surrounding the field work, laboratories, and research centers where scientists work. This was not the GoMRI model. The initiative's leadership was always acutely aware of the need to bring local communities, business, and interest groups into dialogue with scientists and researchers. This would leave a lasting legacy of education and understanding after the program had officially ended in 2020.

Building a Scientific Community: The Lasting Legacy of Oil Spill Science in the Wake of GoMRI

GoMRI's Research Board and management team successfully connected a wide range of stakeholders to the scientific research that its funding generated. This was meant in part to demystify the effects of the Deepwater Horizon oil spill. It was also meant to help relevant user groups prepare to respond to future disasters. By improving the state of knowledge about the health and long-term recovery of the Gulf of Mexico, as well as working to ensure that disasters on the scale of Deepwater Horizon could be addressed quickly and effectively, GoMRI's science served the public interest. GoMRI also provided a significant investment in the ability of the Gulf Coast region to continue educating and training research scientists with expertise and experience relevant to environmental quality research, including effects of oil inputs in the Gulf of Mexico.

The education strategies that GoMRI adopted were diverse. Because of the initiative's distributed management system, the research consortia were largely able to build their own education strategies as they saw fit. This had three advantages. First, having the consortia develop their own education strategies allowed the Research Board to focus on improving key elements of the GoMRI scientific program, such as data management and the RFP process. Second, placing some of the responsibility to develop education strategies on the consortia encouraged the teams to engage with local communities, thereby connecting the Gulf Region to GoMRI-sponsored research. Some consortia worked closely with local teachers. C-IMAGE, for example, brought teachers onto research vessels and allowed them to participate in the scientific process. Those teachers were then connected to a larger network of teacher-participants, who worked together to develop curricula based on their own experience watching and assisting in GoMRI-funded research. Finally, allowing the consortia to develop their own education strategies fostered creativity rather than one standardized approach. This allowed the consortia to adapt to the needs and resources that different communities possessed.[7]

GoMRI's outreach efforts took place in two different ways, according to the individuals responsible for organizing them. Some outreach efforts were directed by the consortia. Later versions of the GoMRI RFP solicitations stipulated that every consortium had to develop an outreach plan, which counted in the evaluation of each proposal. Other outreach efforts were developed by the Research Board's Outreach and Communications Committee and implemented by the GoMRI management team. Those efforts were centralized within the GoMRI management structure, and included efforts undertaken by the initiative as well as its partner organizations. The Smithsonian Ocean Portal, for example, was partner to GoMRI and helped with its national outreach efforts. By providing a clearinghouse for much of the science that GoMRI funds generated, the Smithsonian Ocean Portal provided a gateway for national audiences to access information online about the Deepwater Horizon oil spill and the science that GoMRI supported to better understand its long-term consequences. The Research Board OCC also contracted some elements of its outreach efforts out to other parties. The *Dispatches from the Gulf* series, for example, was filmed, edited, and produced by Screenscope, a documentary production company based in Washington, DC. Hal and Marilyn Weiner, the owners of Screenscope productions, were able to recruit Matt Damon to narrate three hourlong documentaries about the work that GoMRI supported over the course of the 2010s.

Finally, through its capacity building GoMRI has contributed to a lasting interest in and capability for marine environmental research in the Gulf region, prompted by the networks that the Research Board fostered throughout the 2010s. Considered by some scientists to be America's forgotten sea, the Gulf of Mexico often received less attention and funding from the federal government than other major bodies of water, such as the Chesapeake Bay and the Great Lakes. Some of the key figures in GoMRI's history have acknowledged the historical lack of attention paid to the environment, ecology, economy, and communities around the Gulf Coast region and have used their work with GoMRI to redress the region's unequal access to scientific funding.

Through its consortia, GoMRI funded research opportunities for graduate students pursuing both master's degrees and PhDs in oil spill science. It has supported a range of new publications by researchers studying effects of the Deepwater Horizon oil spill. The GoMRI scholars program provided graduate researchers with opportunities to meet members of the Research Board, each other, and other scientists working on GoMRI-funded research, fostering interdisciplinary connections across a range of institutions located in the Gulf of Mexico region.

The annual GoMOSES conferences have done much to support broad, interdisciplinary work. Originating as an annual meeting of the Research Board and GoMRI awardees, over time the conferences came to encompass hundreds of oil spill scientists presenting on a range of research projects across the GoMRI scientific themes. Hosted at different times in all five Gulf Coast states, the GoMOSES conferences have also been open to the public, which can visit and learn about the research being done in various communities. At the convocation of the final GoMOSES conference, Dr. Larry McKinney highlighted the extent to which the Gulf had been a forgotten sea. GoMRI's capacity-building work has helped forge partnerships and connections that will direct much-needed scientific attention to the region. Even after the end of GoMRI in 2020, the Gulf of Mexico Conference, organized by the Gulf of Mexico Alliance, will continue to bring together scientists to discuss the state of the Gulf. Many participants in past GoMOSES conferences are expected to attend. Although it was held virtually in 2021 due to the COVID-19 pandemic, continuation of the GoMRI model for bringing scientists of diverse backgrounds together will continue into subsequent years.

Education, outreach, and capacity building all helped GoMRI work in the interest of the Gulf community. By educating teachers about the scientific research being done on the Gulf of Mexico, GoMRI's consortia developed curricula that connected regional classrooms to the pressing issues that scientists were studying. Outreach programs established dialogues between GoMRI scientists and local communities, ensuring that scientists heard and understood the concerns of local residents. Capacity building bolstered the ranks of researchers equipped to capture data, analyze their findings, and reach conclusions that contributed to the knowledge base of the Gulf region. Ultimately, those efforts will keep the Gulf of Mexico relevant to the oil spill science community.

Translating Knowledge Through Synthesis and Legacy

GoMRI understood early on that the press and the public would be eager to hear what it had learned with $500 million and ten years of research. While GoMRI funding had led to a number of publications across its scientific themes, Colwell and the Research Board could not simply point to highly technical documents as the fruits of GoMRI's labor. The synthesis and legacy efforts were the means by which the initiative translated the scientific knowledge that BP's funds had generated to relevant parties. By conducting synthesis and legacy work during the second half of its operational lifespan, GoMRI answered questions from the Gulf community and ensured that relevant parties had the knowledge necessary to understand, respond to, and mitigate future oil spills.

Under the leadership of the Research Board Synthesis and Legacy Committee, GoMRI developed a comprehensive framework to guide its efforts to translate the knowledge it had generated to relevant user communities. Later synthesis efforts, beginning in April 2017, used five guiding questions and eight core areas to distribute the work of synthesizing oil spill science. From the beginning of its later synthesis efforts, GoMRI strove to incorporate both the knowledge that it had generated through its own funding as well as the knowledge generated from other researchers and institutions. GoMRI's synthesis efforts were intended to clearly articulate the state of oil spill science in various disciplines in 2020, as well as offer new questions for future researchers to investigate in the following years.

The Synthesis and Legacy Committee used a workshop model that it had first introduced in 2012 to conduct synthesis efforts across its eight core areas. Workshop organizers invited interdisciplinary scientists, government and industry representatives, and relevant user communities to contribute to synthesis publications together. While the Research Board offered suggestions for how to structure the synthesis workshops and refereed the process, ultimately the synthesis efforts were led from the bottom up. Research groups were given a fair degree of autonomy in answering the question of what GoMRI had learned.

The synthesis process itself involved attention on the part of researchers to questions of scientific literacy. Synthesis workshop participants were confronted with the difficulty of translating scientific knowledge to nonscientific audiences. Hence, GoMRI and the Consortium for Ocean Leadership invited user communities to participate in the synthesis workshops and voice their questions. Furthermore, GoMRI and its partner organizations had several years of outreach experience that had illuminated the most pressing questions of regional communities. The Research Board understood that simply pointing to publications GoMRI funding had supported would not be useful to relevant user communities. Furthermore, the question of "what did you learn" could be answered in different ways based on which audiences were asking (such as the press, the fishing industry, policymakers). GoMRI's synthesis efforts shed light on the scientific process and what it had revealed over the past ten years.

Lessons Learned: Areas for Future Improvement

The Gulf of Mexico Research Initiative accomplished a great deal during its operational history, producing thousands of high-quality scientific publications, connecting stakeholders to the science its funding generated through outreach and education efforts, and demonstrating how industry funds could be used to build an independent, scientist- and administrator-driven research program. As 2020 approached, members of the Research Board and the management team began to reflect on areas where GoMRI could have strengthened its efforts, if given the chance to do what it had done once again. Acknowledging its successes, the Research Board also emphasized two areas of improvement. This section will describe those two areas and elaborate on the Research Board's comments about them.

The synthesis program that the Research Board implemented was impressive for many reasons. There remains, however, a general consensus among members of the Research Board that synthesis should have started much earlier. In fact, most Research Board members agree that synthesis should have been incorporated into the earliest RFPs. Although synthesis was an item on the agenda at the first management team meeting in 2012, it took another four years to develop into a full-fledged program. In RFP-VI, released in 2016, the Research Board required consortia to develop synthesis plans. In 2017, the Research Board finalized a framework to guide late-stage GoMRI synthesis efforts. Had the Research Board constructed a framework earlier, and had synthesis been a requirement for all consortia seeking funding in earlier RFPs, GoMRI would have been able to conduct synthesis at the same time it was funding research. This would have provided a constant means by which to assess the ways that GoMRI funding was improving the baseline knowledge about oil spill science across its five themes. Regardless, GoMRI synthesis efforts accomplished a lot, given the volume of publications that it produced after the final synthesis framework was adopted in 2017.

The second area of improvement that the Research Board highlighted was in "balancing the bouquet." The metaphor of the balanced bouquet referred to the Research Board's effort to balance funding priorities across its five scientific themes. This proved difficult, and GoMRI had the most trouble allocating money to the fifth scientific theme: "impact of oil spills on public health including behavioral, socioeconomic, environmental risk assessment, community capacity and other population health considerations and issues." In comments that he provided on an initial draft of this manuscript, Dr. Burton Singer, a scientist specializing in public health, pointed out that the fewest consortia proposals were put forward for the public health theme. He attributed this dearth of proposals to the problem of disciplinary boundaries. Public health scientists had little experience collaborating with research groups in other fields, such as marine biology, marine ecology, and oceanography, which received the bulk of GoMRI funding. The Research Board attempted to foster interdisciplinary connections between experts in the field of public health and other scientific fields of study, but, as Singer noted, "by the end of GoMRI, there was only a single consortium focused on community resilience that met the high standards required for GoMRI funding."[8]

The Research Board was able to slightly correct for the lack of emphasis on public health studies in the final two years of GoMRI, when a subset of board members initiated two public health workshops that were "remarkably successful," as Singer notes.[9] Had the Research Board engaged in concerted efforts from the very beginning to bring public health scientists into a program where other fields have a strong history of interdisciplinary collaboration, Singer believes, there might have been more interest within the public health community for the type of research that GoMRI was funding and therefore higher-quality consortia proposals for Theme Five. While the Research Board strove to "balance the bouquet," for various reasons certain themes are more represented among funding awardees than others. Still, the Research Board did its best to bring the public health community into closer collaboration with other researchers.

Providing a succinct history of a $500 million, ten-year endeavor to understand the consequences of a major industrial disaster is no easy task. Necessarily, only part of the story of GoMRI is told here. This manuscript focuses on a few important questions: What was GoMRI's purpose? Its legacy? How did it come to be and who helped build it? What were some of the major obstacles that scientists faced in creating GoMRI? How did they overcome those obstacles?

The answers to those questions illuminate the principles that informed GoMRI's mission. Although GoMRI was established to produce the best possible relevant science, what this meant, as it was practiced, was that the GoMRI model tied scientific research to the needs of the Gulf region. Throughout its decade of funding scientific research, the GoMRI Research Board and management team invested significant time and money in figuring out how to address the concerns of the communities most affected by Deepwater Horizon. By connecting the Gulf region to its research through consortia outreach programs, workshops, panels, conferences, publications, media training, and synthesis workshops, GoMRI made science matter to the first responders, policymakers, industries, and residents of the region. Through its close coordination with partner organizations, GoMRI also presented the knowledge it had generated to national and international audiences. The best possible relevant science, under the GoMRI model, did not exist for scientists alone, far removed from the communities where it was generated. Rather, the best possible relevant science offered new knowledge that could help a region riven by storms, spills, and other disasters move on from the traumas of the past, and better prepare for whatever else might happen in the future.

It stands to mention that the fact that GoMRI was able to conduct independent, transparent, and thorough scientific research was not a given. Colwell and the Research Board built a robust scientific organization that was free from the control and oversight of its original contributor. BP provided $500 million in order to help sponsor scientific research. It did not anticipate the extent to which Colwell would build a truly independent research organization. At first, the company chafed under the arrangement established by the Master

Research Agreement.[10] But, over time, it learned that control over GoMRI rightfully belonged with the scientists and administrators who ran the initiative and were best positioned to foster objective scientific research.

Time will tell how prepared the Gulf Coast is for whatever comes next. Threatened by the long-term effects of climate change, the five Gulf states are preparing for more intense seasonal storms, rising sea levels, and eroding coastlines. If experience in the recent past is an indication, such phenomena may contribute to more oil spills in the region. After a decade of work, it is safe to say that GoMRI has done much to foster research on the Gulf and translate that knowledge to relevant parties. After 2020, it will be up to the Gulf states to put that knowledge to work in disaster preparedness planning. The long-term health and welfare of the Gulf of Mexico and the diverse ecosystems, environments, biota, and communities that it supports depend on it.

APPENDIX A

GULF OF MEXICO RESEARCH INITIATIVE RESEARCH BOARD MEMBERS

Listed below are present and former members of the GoMRI Research Board and years of service. [† deceased]

Chair
Rita Colwell, PhD (2010-2020)
Distinguished University Professor
University of Maryland College Park and
Johns Hopkins University Bloomberg School of Public Health
Director, National Science Foundation (1998-2004)

Vice Chair
Margaret Leinen, PhD (2010-2020)
Director, Scripps Institution of Oceanography at UC San Diego

Debra S. Benoit, MEd (2010-2020)
Director, Research and Sponsored Programs
Nicholls State University

Peter G. Brewer, PhD (2010-2020)
Senior Scientist
Monterey Bay Aquarium Research Institute

Richard E. Dodge, PhD (2010–2020)
Dean
Nova Southeastern University Oceanographic Center

John W. Farrington, PhD (2010–2020)
Dean Emeritus
Woods Hole Oceanographic Institution

Kenneth M. Halanych, PhD (2010–2020)
Alumni Professor and Coordinator, Marine Biology Program
Auburn University

David Halpern, PhD (2010–2020)
Senior Research Scientist, California Institute of Technology/NASA Jet Propulsion Laboratory

William T. Hogarth, PhD[†] **(2010–2020)**
Director, Florida Institute of Oceanography, University of South Florida

Jörg Imberger (2011–2015)
Director of the Centre for Water Research and Professor of Environmental Engineering
University of Western Australia

Cecilie Mauritzen, PhD (2016–2020)
Chief Scientist, Water and Climate
NIVA—Norwegian Institute for Water Research

Raymond L. Orbach, PhD (2010–2020)
Cockrell Family Regents Chair in Engineering
The University of Texas at Austin

Jürgen Rullkötter, PhD (2010-2020)
Professor of Organic Geochemistry (retired)
Institute of Chemistry and Biology of the Marine Environment (ICBM)
University of Oldenburg, Germany

David R. Shaw, PhD (2010-2020)
Vice President for Research and Economic Development
Mississippi State University

Richard F. Shaw, PhD (2013-2020)
Associate Dean, School of the Coast and Environment
Professor, Department of Oceanography and Coastal Sciences
Louisiana State University

John Shepherd, PhD (2010-2020)
Professorial Research Fellow
School of Ocean and Earth Science, National Oceanography Centre
University of Southampton, UK

Bob Shipp, PhD (2010-2020)
Professor Emeritus
University of South Alabama

Burton Singer, PhD (2010-2020)
Adjunct Professor
Emerging Pathogens Institute
University of Florida

Ciro V. Sumaya, MD, MPHTM[†] (2010-2019)
Professor, Health Policy and Management
Cox Endowed Chair in Medicine
Founding Dean, School of Rural Public Health
Texas A&M Health Science Center

Denis Wiesenburg, PhD (2010–2020)
Provost and Vice President
University of Southern Mississippi

Charles A. Wilson, PhD (2010–2012)
Executive Director
University of Louisiana Sea Grant College Program

Dana Yoerger, PhD (2010–2020)
Senior Scientist
Woods Hole Oceanographic Institution

APPENDIX B

GULF OF MEXICO RESEARCH INITIATIVE RESEARCH BOARD COMMITTEES

The names of the Research Board members who served on the committees listed below are confidential.

Annual Meeting Committee
Organized the annual meetings of the Research Board as stipulated in the Master Research Agreement. The meetings came to be a central feature of the annual GoMOSES conferences.

Book Committee
Selected a historian to draft a manuscript on the history of GoMRI.

Budget Committee
Responsible for the financial management of the Gulf of Mexico Research Initiative.

Chief Scientific Officer Selection Committee
Initially led by Research Board member and later CSO Chuck Wilson, this committee evaluated nominees for chief scientific officer.

Data Management Committee
Created GoMRI data management policies based on requirements of the Master Research Agreement. Recruited appropriate institutions and individuals to store and maintain data and metadata generated with GoMRI funding.

Ethics Committee
Ensured that the GoMRI Research Board, management team, and funding recipients adhered to the GoMRI Master Research Agreement, by-laws, ethics policy, and code of conduct.

Governance Committee
Organized Gulf of Mexico Research Initiative leadership structure to ensure compliance with the Master Research Agreement.

Outreach and Communications Committee
Oversaw implementation of outreach and communication strategies of the GoMRI management team. Worked with Sea Grant, Smithsonian Institute, and Ocean Leadership to disseminate research generated with GoMRI funding.

RFP Committees (RFP I-RFP VI)
Drafted language of each RFP. Worked with Ocean Leadership to advertise RFPs to colleges, universities, and research institutions. Developed requirements for RFP submissions. Established RFP review process.

Synthesis and Legacy Committee
Developed GoMRI legacy goals and process for conducting synthesis, organizing synthesis, and publication of synthesis results.

APPENDIX C

GULF OF MEXICO RESEARCH INITIATIVE CONSORTIA

Listed below are all of the consortia that GoMRI funded and their years of operation.

Aggregation and Degradation of Dispersants and Oil by Microbial Expolymers (ADDOMEx) [*2015-2020*]

Alabama Center for Ecological Resilience (ACER) [*2015-2020*]

Center for Integrated Modeling and Analysis of Gulf Ecosystems (C-IMAGE) [*2011-2020*]

Coastal Waters Consortium (CWC) [*2011-2020*]

Consortium for Advanced Research on Marine Mammal Health Assessment (CARMMHA) [*2018-2020*]

Consortium for Advanced Research on Transport of Hydrocarbon in the Environment (CARTHE) [*2011-2020*]

Consortium for the Molecular Engineering of Dispersant Systems (C-MEDS) [*2011–2015*]

Consortium for Oil Spill Exposure Pathways in Coastal River-Dominated Ecosystems (CONCORDE) [*2015–2020*]

Consortium for Resilient Gulf Communities (CRGC) [*2015–2020*]

Consortium for Simulation of Oil-Microbial Interactions in the Ocean (CSOMIO) [*2018–2020*]

Deep-Pelagic Nekton Dynamics Consortium (DEEPEND) [*2015–2020*]

Deepsea to Coast Connectivity in the Eastern Gulf of Mexico (DEEP-C) [*2011–2015*]

Dispersion Research on Oil: Physics and Plankton Studies (DROPPS) [*2011–2020*]

Ecosystem Impacts of Oil and Gas Inputs to the Gulf (ECOGIG) [*2011–2019*]

Gulf of Mexico Integrated Spill Response Consortium (GISR) [*2011–2016*]

Littoral Acoustic Demonstration Center—Gulf Ecological Monitoring and Modeling (LADC-GEMM) [*2015–2019*]

Relationship of Effects of Cardiac Outcomes in Fish for Validation of Ecological Risk (RECOVER) [*2015–2020*]

APPENDIX D

GULF OF MEXICO RESEARCH INITIATIVE MANAGEMENT TEAM MEMBERS

Listed below are the members of the GoMRI management team and their period of affiliation with GoMRI.

American Institute of Biological Sciences
DaJoie Croslan (2010–2013)
Arati Deshmukh (2016–2020)
Scott Glisson (2010–2021)
Robert Gropp (2010–2020)
Syreeta Jones (2010–2021)
Caitlin McPartland (2012–2018)
Jennifer Petitt (2010–2021)
Joanne Sullivan (2016–2020)

Consortium for Ocean Leadership
Michael Feldman (2015–2021)
Katie Fillingham (2015–2020)
Suzanne Garrett (2015–2020)
Megan Gibney (2013–2014)
Jenny Hauser (2012–2015)
Heather Mannix (2011–2014)

Jessie Swanseen (2012–2021)
Callan Yanoff (2018–2021)
Kristen Yarincik (2011–2021)
Leigh Zimmermann (2011–2021)

Gulf of Mexico Alliance
Brittany Ballew (2015–2017)
Mike Carron (2010–2021)
Chris Kirby (2011–2020)
Kevin Shaw (2011–2021)
Devany Tyler (2012–2020)
Chuck Wilson (2012–2021)*

Harte Research Institute for Gulf of Mexico Studies
Sandra Ellis (2013–2018)
Jim Gibeaut (2012–2021)
Rosalie Rossi (2015–2021)
Lauren Showalter (2012–2016)

Northern Gulf Institute
Maggie Dannreuther (2010–2021)
Stephanie Ellis (2014–2021)
Jarryl Ritchie (2010–2021)
Suzanne Shean (2010–2021)

Sea Grant
Dani Bailey (2020–2021)
Larissa Graham (2014–2017)
Chris Hale (2014–2019)
Emily Maung-Douglass (2014–2021)
Missy Partyka (2018–2021)
Stephen Sempier (2014–2021)
Tara Skelton (2016–2021)

LaDon Swann (2014–2021)
Monica Wilson (2014–2021)

*Wilson served as a member of the Research Board from 2010–2011 before being appointed GoMRI chief scientific officer.

APPENDIX E

REQUEST FOR PROPOSAL TIMELINE[1]

April 2010
The Deepwater Horizon oil spill begins. The well would not be capped until the following September.

May 2010
BP announces its commitment to provide $500 million in research funding over the following ten years.

June 2010
Year One Block Grants
$45 million in funding provided directly from BP to Gulf state institutions and the National Institutes of Health to establish critical baseline data as the foundation for subsequent research as well as support for studying the health of the oil spill workers and volunteers.

July 2011
RFP-III Bridge Grants
$1.5 million in grants to seventeen projects supporting continuity of observations and sampling while the peer-review process was underway for Year 2–4 Consortia.

August 2012
RFP-I Consortia Grants (Years 2–4)
$110 million in grants to eight research consortia comprising experts from over six dozen research institutions in twenty-seven US states and five countries.

August 2013
RFP-II Investigator Grants (Years 3–5)
$18.5 million in grants to nineteen individuals or collaborative efforts involving a principal investigator (PI) and up to three co-principal investigators (co-PIs).

January 2015
RFP-IV Consortia Grants (Years 5–7)
$140 million in grants to twelve research consortia.

January 2016
RFP-V Investigator Grants (Years 6–8)
$38 million in grants awarded to individuals and teams studying the effects of oil on the Gulf of Mexico ecosystem and public health. A total of twenty-two research proposals were funded under this GoMRI program.

January 2018
RFP-VI Research Grants (Years 8–10)
$50 million in grants to support research into effects of the Deepwater Horizon incident on the Gulf of Mexico ecosystem. A total of thirty-one research proposals—eight Research Consortia and twenty-three small research teams.

TOTAL: $403 million

APPENDIX F

GULF OF MEXICO RESEARCH INITIATIVE BY-LAWS

BYLAWS (APPROVED OCTOBER 18, 2013)

I. Preamble

The Gulf of Mexico Research Initiative Research Board (the Board) is established by and subject to the latest version of the Master Research Agreement (MRA), between BP Exploration and Production Inc. (BP) and the Gulf of Mexico Alliance (the Alliance)—together referred to as "the Parties"—to create and fund an independent research program, to be known as the Gulf of Mexico Research Initiative (GoMRI), to study the effects, and the potential associated impacts, of hydrocarbon releases on the environment and public health in the Gulf of Mexico, as well as to develop improved spill mitigation, oil detection, characterization and remediation technologies.

II. Purpose

The Board serves as the decision-making and oversight body regarding scientific research and other activities funded by and conducted through the GoMRI. The Board has the responsibility and authority to ensure the intellectual quality, research effectiveness, and independence of the GoMRI programs; to select the Research Consortia (groups of nongovernmental or nonprofit academic and research institutions) and smaller groups of individual investigator projects that are to receive GoMRI funds for approved research projects pursuant to merit review by peer evaluation as described in the 2005 National Science Board report on the Peer Evaluation Process (NSB-05-119), and to perform an annual review and approval for funding of all research projects.

III. Governance: Core Adherence to Master Research Agreement

The Board shall operate and govern itself in compliance with the MRA and that document's Section 3 ("Research Board"). Should any inconsistencies arise between these Bylaws and the MRA, the MRA shall have precedence.

With reference to the numbered sections of the MRA as amended July 11, 2012, the following apply for the conduct of the Board:

General Role and Responsibilities: Sections 3.1 and 3.2
Internal Operations: Section 3.3
Composition: Section 3.4
Removal / Resignation: Section 3.5
Chair: Section 3.6
Procedural Matters (incl. voting, quorums, etc.): Section 3.7
Standards of Conduct (incl. Conflict Of Interest): Section 3.8 and Appendix 2
Chief Scientific Officer: Section 3.9
Administrative Support: Section 3.10
Debriefings: Section 3.11
Indemnification: Section 3.12
Certification: Section 3.13
Matters Reserved to the Parties: Section 3.14

IV. Membership

a. Appointments

There shall be two pools of Board Members initially appointed by Gulf of Mexico Alliance (GOMA) and BP.

1. GOMA Board Members: at all times there will be two members appointed by GOMA from each of the five Gulf of Mexico states (Alabama, Florida, Louisiana, Mississippi, and Texas).
2. BP Board Members: at all times there will be 10 members appointed by BP from the at-large scientific and research communities.

b. Board Member Participation / Removal or Resignation

Board members who are unable to complete their work in a timely fashion may have their duties re-assigned by the chair in order that the Board continues to meet its responsibilities to the GoMRI, in particular the project timeline in Section

3.2.9 of the MRA. Two occurrences of such non-participation on a Board member's part, or a violation of the Board's Code of Conduct, or the failure of a Board member to attend two consecutive Board meetings without the prior approval of the chair for their absence, shall initiate provisions in the MRA in which the Board recommends to the member that he or she resign pursuant to Section 3.5.2, or the Board recommends to the Party that appointed such member that the member be removed pursuant to Section 3.5.1.

c. Board Member Replacement

Should a Board member vacate his/her seat the appropriate appointing authority will be notified.

d. Qualifications

All Board member replacements will meet the following qualifications and be an expert in a science discipline directly related to the work of GoMRI, including public health.

- Hold an advanced degree in their field.
- Hold or have held a full-time or emeritus faculty-level position at an academic institution, a research institution, or a federal government science agency.
- Must not currently receive GoMRI funding or if currently receiving will give up GoMRI funding on becoming a Board member.

All Research Board member appointments will be consistent with MRA Section 3.4.

V. Additional Provisions

a. Code of Conduct

The Board shall, in all regards, adhere to its Code of Conduct, attached as Appendix 1.

b. Term of the Chair

As per Section 3.6 of the MRA ("Chairman"), the Chair shall be a member of the Board. Dr Rita Colwell will be the inaugural Chair. Subsequent Chairs shall be elected by the Board to serve in three-year terms on a calendar year basis. In the year before the expiration of a Chair's term, a Nominating Committee consisting of the Vice Chair (or another RB member appointed by the Chair if the Vice

Chair wishes to be considered), the Chief Scientific Officer, and two other members of the Board as appointed by the Chair—one from the set of Board members appointed by BP and one from those appointed by the Alliance—shall determine if the current Chair wishes to run for re-election (multiple terms are permitted), shall identify any other candidates there might be on the Board for Chair, then shall bring the slate to the Board to vote on as per Section 3.7 of the MRA ("Procedural Matters").

c. Term of the Vice Chair

The Vice Chair is appointed by and serves at the pleasure of the sitting Chair.

d. Committees

Committee members must be members of the Board. The Chair shall appoint and charge the chairs and members of such committees as may be deemed necessary for the operations of the Board and the success of the GoMRI. Vacancies on such committees shall be filled by the Chair. The CSO shall be an ex-officio member of all committees.

e. Amendments

Amendments to these Bylaws may be initiated by any member of the Board and must be approved by a two-thirds majority vote of the Board as per Section 3.7 of the MRA ("Procedural Matters").

— END —

APPENDIX G

GULF OF MEXICO RESEARCH INITIATIVE RESEARCH BOARD CODE OF CONDUCT

CODE OF CONDUCT

Appendix 1 to Bylaws: Code of Conduct
Approved by GoMRI RB Governance Committee on February 18, 2015
Approved by GoMRI RB on February 19, 2015

Gulf of Mexico Research Initiative (GoMRI) Research Board (RB)
GoMRI Management Team (MT) and RB Chief Scientific Officer

PREAMBLE

- Purpose for serving on the Gulf of Mexico Research Initiative (GoMRI) Research Board (RB) and GoMRI Management Team (MT) is the fulfillment of the objectives of GoMRI, including
- Commitment to generate new scientific knowledge and improved understanding of the predictability of the impact of oil spills in the Gulf of Mexico and elsewhere.
- Commitment to highest integrity stewardship of GoMRI funds.

- The RB consists of twenty people, ten appointed by the Gulf of Mexico Alliance (GOMA) and ten appointed by BP plc.
- Unless stated otherwise in the BP-GOMA Master Research Agreement (MRA), the RB shall serve as the decision-making and oversight body regarding the research conducted pursuant to the GoMRI.

- The MRA gives autonomy to the RB concerning selection of proposals submitted in response to the RB Requests for Proposals (RFPs) and other relevant Announcements of Opportunity (AOs).
- The MRA stipulates that neither BP nor GOMA may seek to influence the activities of the RB.
- RB members are expected to serve as individual scientists and not to represent any organization.
- The RB is an independent advisory group and, thus, does not represent any constituency, stakeholder, or interest group within the governmental, nongovernmental, and private sectors.
- The RB Chief Scientific Officer (RBCSO), who is affiliated with GOMA, is responsible for coordinating science activities under the direction of the RB.
- The MT consists of:
    - GoMRI Administrative Unit, which is composed of personnel affiliated with GOMA and primary subcontractors from the Consortium for Ocean Leadership (COL), the Harte Research Institute for Gulf of Mexico Studies, and the Northern Gulf Institute
    - GoMRI Grants Unit, which is composed of personnel affiliated with COL and GOMA
    - RB Administrative Entity, which is composed of personnel affiliated with the American Institute of Biological Sciences
    - RBCSO
    - Additional contractors supporting the work of the MT on RB activities.

CODE OF CONDUCT

- Each member of the RB and MT associated with the workings of the RB will:
    - Operate with the highest standards of personal integrity and professional scientific objectivity.
    - Select the highest quality science and most appropriate investment proposals submitted in response to RFPs and AOs.
    - Adhere to decision procedures that follow the practice of merit review by peer evaluation, as described in the 2005 Report of the National Science Board (NSB-05-119), or any update thereto.
    - Make every effort to meet the workload expectations of the RB, including conference calls, RB face-to-face meetings, proposal

reviews and research project reviews, and other duties as assigned by the Chair.

- A real or potential conflict of interest by a RB member or by a MT member supporting the work of the RB shall be managed in compliance with the practices of the National Science Foundation, including the practices specified in Form 1230P (2/04) "Conflict of Interests and Confidentiality Statement for NSF Panelists," included as Appendix 2 in the MRA and the RB conflict of interest policy document (https://gulfresearch initiative.org/gri-research-board/governance/conflict-of-interest/).

- Each RB member and MT member supporting RB activities shall:
    - Recuse him/herself from advising or assisting individuals, groups, or institutions with preparation of a proposal to be submitted to the RB.
    - Recuse him/herself during discussions of proposals in which they have a real or potential conflict of interest and for RB members, in which they have a real or potential conflict of interest, from voting on individual proposals submitted to the RB in accordance with the National Science Foundation Board policy (Manual #50). If a real or potential conflict of interest is raised, it will be resolved according to the RB conflict of interest policy document (https://gulfresearch initiative.org/gri-research-board/governance/conflict-of-interest/).
    - Not benefit financially from any decision of the RB other than acceptance of an annual honorarium for duties accomplished as a RB member and wages for a MT member associated with the workings of the RB.

- All communications and discussions, including, but not limited to, face-to-face meetings, teleconferences, emails and faxes, by the RB and between the RB and MT will be confidential.

- There will be an embargo on release of information from the RB and MT before it is made public; this includes policies, RFPs, AOs, etc.

- The RB Chair will represent the RB to the press, media and other organizations. Members of the RB, MT and RBCSO will not discuss the activities of the RB without consent of the Chair.

- The RB and MT are subject to the requirements of the MRA. To the extent that any provision of this Code of Conduct conflicts with the MRA, the MRA shall govern.

NOTES

INTRODUCTION: HOW SCIENTISTS RESPONDED TO A MAJOR OIL SPILL: BUILDING A RESEARCH ORGANIZATION THAT ADDRESSES A SOCIETAL NEED

1. Examples of public institutions include federal, state, and municipal departments and laboratories.

2. Smithsonian Institution, Smithsonian Ocean Portal, Gulf Oil Spill. "The Spill: The Oil's Spread," https://ocean.si.edu/conservation/pollution/gulf-oil-spill#:~:text=The%20Oil's%20Spread,-Mark%20Dodd%2C%20a&text=Over%20the%20course%20of%2087,accidental%20ocean%20spill%20in%20history (accessed August 15, 2024); National Oceanic and Atmospheric Administration, "Oil Spills," https://www.noaa.gov/education/resource-collections/ocean-coasts/oil-spills (accessed August 16, 2024).

3. This is not to imply that other major spill responses were not coordinated or properly funded. The *Exxon Valdez* Oil Spill Trustee Council, formed in 1991, still releases annual financial reports detailing its work in the areas of "Restoration, Habitat, and Administration." See: *Exxon Valdez* Oil Spill Trustee Council, "2020 Annual Financial Report on the EVOSTC Fiscal Year 2019," https://evostc.state.ak.us/media/7305/2020-annual-report-on-the-evostc-fy19-rev-07142020.pdf (accessed August 16, 2024). The trustee council consists of six representatives from government agencies and departments; three are appointed by the federal government and three by the State of Alaska. The work initiated by the trustee council was funded by the $900 million civil settlement between Exxon, the federal government, and the State of Alaska.

4. GoMRI-funded research covered five broad theme areas, which are described in chapter one. They can also be found here: "General FAQs: Frequently Asked Questions," https://gulfresearchinitiative.org/about-gomri/faqs/ (accessed August 16, 2024).

5. GoMRI's Research Board was its chief leadership body. Its formation is discussed in chapters one and two.

6. Interestingly, the virtual office structure that GoMRI relied on preceded the COVID-19 pandemic, which, by March 2020, led to the virtualization of business in general.

7. The nature of GoMRI's research consortia—their formation, operation, and purpose—is discussed in detail in chapters two and four.

8. The most recent synthesis of scientific knowledge about the *Exxon Valdez* spill was published in May 2018. See Kris Holderied and Donna Aderhold, "Science Coordination and Synthesis for the Long-Term Monitoring Program," Exxon Valdez *Oil Spill Trustee Council Project* 16120114-H *Final Report*, Exxon Valdez, *Oil Spill Long-term Monitoring Program (Gulf Watch Alaska) Final Report*, May 2018, http://gulfwatchalaska.org/wp-content/uploads/2018/08/16120114-H-Holderied-and-Aderhold-2018-Final-Report.pdf (accessed August 16, 2024).

9. These numbers are taken from the GoMRI dashboard, compiled by Jay Ritchie, available at https://gomri.org/admin-dashboard/ (accessed July 31, 2024). It helpfully breaks down the number of publications by RFP, scientific theme, and other categories.

CHAPTER ONE: DEEPWATER HORIZON AND THE ORIGIN OF THE GULF OF MEXICO RESEARCH INITIATIVE

1. Daniel Jacobs, *BP Blowout: Inside the Gulf Oil Disaster* (Washington, DC: Brookings Institution Press, 2016), vii.

2. David Barstow, with reporting by Barstow, David Rohde, and Stephanie Saul, "Deepwater Horizon's Final Hours: Missed Signals. Indecision. Failed Defenses. Acts of Valor," *New York Times*, December 26, 2010.

3. Barstow, "Deepwater Horizon's Final Hours"; Jacobs, 1; Ian Urbina, "At Issue in Gulf: Who Was in Charge? Hodgepodge of Oversight for Rig Helped Set Stage for Disaster," *New York Times*, June 6, 2010.

4. Conversation with Dr. Ellen Williams, March 12, 2020.

5. BP Press Office, "Former TNK-BP Head Dudley Gets BP Board Seat," February 18, 2009; available here: https://www.reuters.com/article/business/energy/former-tnk-bp-head-dudley-gets-bp-board-seat-idUSLI660337/ (accessed August 16, 2024); GoMOSES Opening Symposium, "Broader Horizons: Reflections from Rita Colwell and Ellen Williams," February 3, 2020.

6. Terry Macalister and Richard Wray, "Tony Hayward to Quit BP," *The Guardian*, July 26, 2010; available here: https://www.theguardian.com/business/2010/jul/26/tony-hayward-to-quit-bp (accessed August 16, 2024).

7. This is based on information available in the NSF's grant award search engine, which can be found here: https://www.nsf.gov/awardsearch/simpleSearchResult?queryText=RAPID+%22Deepwater+Horizon%22 (accessed August 16, 2024).

8. Comments on draft manuscript by Dr. Peter Brewer, May 17, 2022.

9. National Oceanic and Atmospheric Administration, Damage Assessment, Remediation, and Restoration Program, Deepwater Horizon, https://darrp.noaa.gov/oil-spills/deepwater-horizon (accessed August 16, 2024).

10. Office of Science and Technology Policy, "Scientific Research on the Effects and Fate of Oil in the Ocean," meeting invitation, May 19, 2010. Provided to the author by Dr. Ellen Williams.

11. Office of Science and Technology Policy, "Scientific Research on the Effects and Fate of Oil in the Ocean."

12. BP Press Office, "BP Pledges $500 Million for Independent Research into Impact of Spill on Marine Environment," May 24, 2010; available here: https://www.bp.com/en/global/corporate/news-and-insights/press-releases/bp-pledges-500-million-for-independent-research-into-impact-of-spill-on-marine-environment.html (accessed August 16, 2024).

13. National Science Foundation website, "Office of the Director (OD)," https://nsf.gov/od/ (accessed August 16, 2024).

14. Sea Grant, webpage, "Sea Grant is a Federal-University partnership program that brings science together with communities for solutions that work," https://seagrant.noaa.gov/About (accessed August 16, 2024).

15. Conversation with Dr. Rita Colwell, March 25, 2021.

16. National Science Board website, "About the NSB: Background," https://www.nsf.gov/nsb/about/index.jsp (accessed August 16, 2024).

17. National Science Board website, "About the NSB: Background."

18. Conversation with Dr. Ellen Williams, March 12, 2020.

19. Letter from Dr. Ellen Williams to Dr. Rita Colwell, May 26, 2010. Provided to the author by Dr. Ellen Williams.

20. The development of the board is discussed in chapter two.

21. BP Press Office, "Announcement of Independent Advisory Council and Request for Proposals for Gulf of Mexico Research Initiative," June 12, 2010. Provided by Dr. Ellen Williams to the author.

22. Amanda Mascarelli, "White House Stalls Oil-Slick Research," *Nature* 465 (June 24, 2010), 993.

23. Mascarelli, "White House Stalls Oil-Slick Research."

24. Technically, this book is funded by the Gulf of Mexico Alliance. GoMRI refers to the Research Board and Management Team as well as the set of partnerships between various organizations, detailed below, that received funding in part from the Gulf of Mexico Research Initiative. The entity that directs payments and contracts with researchers is the Gulf of Mexico Alliance.

25. US Commission on Ocean Policy, *An Ocean Blueprint for the 21st Century* (Washington, DC: Government Printing Office, 2004), 1.

26. The Commission was chaired by Admiral James D. Watkins, chairman and president emeritus at the Consortium for Oceanographic Research and Education; Dr. Robert Ballard, Professor of Oceanography; Ted A. Beattie, president and chief

executive officer of the John G. Shedd Aquarium; Lillian Borrone, former assistant executive director for the Port Authority of New York and New Jersey; James M. Coleman, Boyd Professor at the Coastal Studies Institute of Louisiana State University; Ann D'Amato, chief of staff to the City Attorney of Los Angeles; Lawrence Dickerson, president and chief operating officer of Diamond Offshore Drilling, Inc.; Vice Admiral Paul G. Gaffney II, president of Monmouth University; Marc J. Hershman, professor at the School of Marine Affairs at the University of Washington; Paul L. Kelley, senior vice president of Rowan Companies, Inc.; Christopher Koch, president and chief executive officer of the World Shipping Council; Frank Muller-Karger, PhD, professor at the College of Marine Science at the University of South Florida; Edward B. Rasmuson, chairman of the board of directors of Wells Fargo Bank; Dr. Andrew A. Rosenberg, professor, Department of Natural Resources and Institute for the Study of Earth, Oceans, and Space, University of New Hampshire; William D. Ruckelshaus, strategic director, Madrona Venture Group and the first administrator of the Environmental Protection Agency; Dr. Paul A. Sandifer, senior scientist at the National Oceanic and Atmospheric Administration; and Dr. Thomas Kitsos, executive director of the Commission.

27. An Act to establish a Commission on Ocean Policy, and for other purposes, Public Law 106–256. U.S. Statues at Large 114 (2000): 644, https://www.govinfo.gov/content/pkg/STATUTE-114/pdf/STATUTE-114-Pg644.pdf (accessed August 16, 2024).

28. US Commission on Ocean Policy (from note 34), 86.

29. US Commission on Ocean Policy, 88–89, 93.

30. Gulf of Mexico Alliance website, "Alliance History," https://gulfofmexicoalliance.org/about-us/alliance-timeline/ (accessed August 16, 2024).

31. Email with Laura Bowie, executive director of the Gulf of Mexico Alliance, June 9, 2021.

32. Gulf Mexico Alliance website, "Organization and Partners," https://gulfofmexicoalliance.org/about-us/organization-and-partnerships/ (accessed August 16, 2024).

33. Gulf Mexico Alliance website, "Organization and Partners."

34. Gulf of Mexico Alliance website, "Gulf Star," https://gulfofmexicoalliance.org/gulf-star/ (accessed August 16, 2024). The Gulf Star Partnership program is a public-private partnership administered by the Gulf of Mexico Alliance. As a 501(c)(3) public charity, the Gulf Star Partnership program was able to accept tax-deductible contributions from individuals, businesses, and other private organizations to fund research and landscape engineering projects in the Gulf of Mexico.

35. Conversation with Kevin Shaw, March 8, 2019. Travel expenses for Research Board members and some other financial items were paid through the AIBS in order to maintain independence from BP. This is discussed in greater detail in chapter two.

36. Conversation with Kevin Shaw, March 8, 2019.

37. Gulf of Mexico Research Initiative, *Master Research Agreement Dated March 14th, 2011 between BP Exploration & Production, Inc. and Gulf of Mexico Alliance as Amended and Restated December 1st, 2011*, available at https://gulfresearchinitiative.org/wp-content/uploads/2011/04/Amended-and-Restated-GoMRI-BP-Gulf-of-Mexico-Master-Research-Agreement-w_-Appendix-3-Final-V1-1-Dec1-2011.pdf (accessed August 16, 2024), 10.

38. Gulf of Mexico Research Initiative, *Master Research Agreement*.

39. Interview with Colwell, March 25, 2021.

40. Interview with Leinen, March 24, 2021.

41. Interview with Leinen, March 24, 2021.

42. Gulf of Mexico Research Initiative website, "Gulf of Mexico Research Initiative Names Chief Scientific Officer," http://gulfresearchinitiative.org/gulf-of-mexico-research-initiative-names-chief-science-officer/ (accessed August 16, 2024).

43. Interview with Colwell, March 25, 2021.

44. Interview with Colwell, March 25, 2021.

45. GoMRI morning Research Board meeting at the GoMOSES Conference, February 4, 2019.

46. William R. Freudenberg and Robert Gramling, *Blowout in the Gulf: The BP Oil Spill Disaster and the Future of Energy in America* (Boston: MIT Press, 2011), 13.

47. These figures and dates can be found in Freudenberg and Gramling, *Blowout in the Gulf*, 12–13.

48. When drilling an oil well, drilling fluid is pumped into the well to balance the pressure in the borehole against the pressure exerted by the surrounding rock. When the pressure exerted by the surrounding rock is greater than that in the borehole, the fluids and gasses from the surrounding rock can enter the borehole, causing the drilling equipment and well to suddenly shudder. This is known as "kicking." For the depth of the Macondo well, see "Committee on the Analysis of Causes of the *Deepwater Horizon* Explosion, Fire, and Oil Spill to Identify Measures to Prevent Similar Accidents in the Future," *Macondo Well Deepwater Horizon Blowout: Lessons for Improving Offshore Drilling Safety* (Washington, DC: National Academies Press, 2012), 4.

49. Arne Jernelöv and Olof Lindén, "Ixtoc I: A Case Study of the World's Largest Oil Spill," *Ambio* 10, no. 6 (1981), 299; 299–306.

50. John Farrington, "Written Testimony of John W. Farrington, Interim Dean and Professor, School of Marine Science and Technology, University of Massachusetts-Dartmouth, and Scientist Emeritus, Woods Hole Oceanographic Institution," Testimony to the National Commission for the BP Deepwater Horizon Oil Spill and Offshore Drilling, Corrected Version, October 1, 2010, 2.

51. Farrington, "Written Testimony."

52. Farrington, "Written Testimony," see 2, 9.

53. Farrington, "Written Testimony," 3.

54. Farrington, "Written Testimony."

55. Conversation with Dr. Steve Murawski, February 5, 2019.

56. Vinnem is quoted in Robert Campbell, "BP's Gulf Battle Echoes Monster '79 Mexico Oil Spill," *Reuters*, May 24, 2010, https://www.reuters.com/article/us-oil-rig-mexico-sidebar/bps-gulf-battle-echoes-monster-79-mexico-oil-spill-idUSTRE64N57U20100524 (accessed August 16, 2024).

57. Campbell, "BP's Gulf Battle Echoes Monster '79 Mexico Oil Spill."

58. Vanessa Romo, "Oil Spill Seeping into Gulf of Mexico Contained after 14 Years, Coast Guard Says," *NPR*, May 16, 2019, https://www.npr.org/2019/05/16/724164873/oil-spill-seeping-into-gulf-of-mexico-contained-after-14-years-coast-guard-says (accessed August 16, 2024).

59. Teresa Sabol Spezio, *Slick Policy: Environmental and Science Policy in the Aftermath of the Santa Barbara Oil Spill* (Pittsburgh: University of Pittsburgh Press, 2018), 1.

60. John Farrington, "Keynote Address," Synthesis Symposium, February 3, 2020.

61. The American Petroleum Institute did conduct research that was not readily available to nonindustry scientists. Information obtained from a conversation with Dr. Chuck Wilson and Dr. Mike Carron, September 18, 2024.

62. National Commission on the BP Deepwater Horizon Oil Spill and Offshore Drilling, 143.

63. National Commission on the BP Deepwater Horizon Oil Spill and Offshore Drilling, 144.

64. Rosalie R. Rossi, Deborah A. LeBel, and James Gibeaut, "Growing Pains of a Data Repository: GRIIDC's Evolution from Environmental Disaster Rapid Response to Promoting FAIR Data," *Frontiers in Climate* 4 (August 24, 2022): 1; https://doi.org/10.3389/fclim.2022.958533.

65. Rossi, LeBel, and Gibeaut, "Growing Pains of a Data Repository."

66. Freudenberg and Gramling, 12.

67. Bryan Walsh, "Oil Spill: Goodbye Mr. Hayward," Time.com, July 25, 2010, http://science.time.com/2010/07/25/oil-spill-goodbye-mr-hayward/ (accessed June 4, 2019).

68. Walsh, "Oil Spill: Goodbye Mr. Hayward."

69. Pete Spotts, "Gulf Oil Spill: BP Grants $500 Million for Independent Research," *Christian Science Monitor*, July 3, 2010; https://www.csmonitor.com/Environment/2010/0703/Gulf-oil-spill-BP-grants-500-million-for-independent-research (accessed August 16, 2024).

70. Spotts, "Gulf Oil Spill."

71. Spotts, "Gulf Oil Spill."

72. Spotts, "Gulf Oil Spill."

CHAPTER TWO: DISTRIBUTED MANAGEMENT: ADMINISTERING SCIENTIFIC RESEARCH DURING AN UNFOLDING CRISIS

1. Email to author, May 15, 2024.

2. The Pew Research Center shows that at the end of 2005, 36 percent of adults in the United States had access to home broadband connections. At the close of 2010, that number increased to 62 percent. In February 2019, nearly three-quarters of all Americans had home broadband connections. See Pew Research Center, "Internet/Broadband Fact Sheet," June 12, 2019, https://www.pewresearch.org/internet/fact-sheet/internet-broadband/ (accessed August 16, 2024). Broadband internet access became a vital tool as the COVID-19 pandemic initiated a wave of lockdowns in March 2020. Reliable internet access became a necessity as many office workers began extended periods of working from home.

3. More on this is found in the following chapter, which covers GRIIDC more specifically.

4. While this manuscript was being prepared, COVID-19 rendered it impossible for in-person meetings to occur safely across the United States. Still, GoMRI required little adjustment in its standard operations because most of its meetings were already virtual. The GoMRI Management Team purposely employed virtual meeting tools to maintain continuity of operations among several different partner organizations scattered across the Gulf region and the United States. While campuses and institutions across the country shuttered throughout most of 2020, GoMRI was insulated from interruption because its staff and partners already functioned at a physical distance through virtual communication.

5. Individuals who were involved with GoMRI from the start frequently describe the early days of GoMRI as "building the plane as it was taking off." With oil continuing to leak from the broken wellhead and $500 million from BP secured, leadership within the initiative had to act quickly to draft, circulate, review, and approve the first research grants. This was a necessary step toward getting the plane off the ground.

6. Interview with Colwell, March 25, 2021. Dr. Leinen's professional biography can also be found here, on the Scripps Institution of Oceanography website: https://mleinen.scrippsprofiles.ucsd.edu/bio/ (accessed August 16, 2024).

7. Gulf of Mexico Research Initiative, *Master Research Agreement Dated March 14th, 2011 between BP Exploration & Production, Inc. and Gulf of Mexico Alliance as Amended and Restated December 1st, 2011.*

8. Gulf of Mexico Research Initiative, *Master Research Agreement*. According to the by-laws, during the calendar year preceding the expiration of a chair's term, a nominating committee would be established to oversee the election process. The nominating committee consisted of the vice chair (or another Research Board member if the vice chair did not wish to be nominated), the chief scientific officer, and two board members appointed by the chair (one from the set appointed by BP

and the other from the set appointed by GOMA). The nominating committee was responsible for determining if the current chair wished to run again, identifying other candidates who wished to be nominated, and bringing the slate of candidates to a vote. The chair election process provided accountability for the GoMRI leadership. Ultimately, the board was confident in Colwell's leadership and no alternative candidates came forward to run for the position of chair.

9. Conversation with Wilson and Carron, January 7, 2021.

10. Conversation with Wilson and Carron, January 7, 2021.

11. See Appendix A.

12. Gulf of Mexico Research Initiative, "Appendix 1 to By-Laws," *Code of Conduct*, February 19, 2015, https://gulfresearchinitiative.org/gri-research-board/governance/code-of-conduct/ (accessed August 16, 2024).

13. Interview with Colwell and Leinen.

14. A list of all of the committees that fall under the Research Board's control can be found in Appendix B.

15. Gulf of Mexico Research Initiative, *Master Research Agreement between BP Exploration and Production, Inc., and Gulf of Mexico Alliance as amended and restated December 1st, 2011*, 16; Memorandum from Chuck Wilson for the GoMRI Research Board Data Committee to Gulf of Mexico Research Initiative (GoMRI) RFP-VI Research Consortia and Individual Investigators, "Update of GoMRI Data Policy and Procedures," August 29, 2017, available at https://gulfresearchinitiative.org/wp-content/uploads/2017/09/Update-of-GoMRI-Data-Policy-and-Procedures-Memorandum-August-29-2017.pdf (accessed August 16, 2024).

16. Gulf of Mexico Research Initiative, "By Laws," webpage.

17. Interview with Wilson, March 13, 2019.

18. Gulf of Mexico Research Initiative press release, "Gulf of Mexico Research Initiative Names Chief Scientific Officer."

19. GoMOSES is discussed in greater detail in chapter four.

20. Interview with Kevin Shaw, February 6, 2019.

21. Interview with Kevin Shaw, March 8, 2019.

22. Gulfbase, "Dr. Mike Carron," https://www.gulfbase.org/people/dr-mike-carron (accessed August 16, 2024).

23. *Ex officio* Research Board members were nonvoting participants in board meetings. They were privy to most of the proceedings of the full board but were barred from the meeting room when executive sessions were conducted.

24. For comparison, the NSF's 2018 budget request to Congress would have provided 81 percent of its total funds for "Research and Related Activities." The other 19 percent would be applied to "Education and Human Resources," "Major Research Equipment and Facilities Construction," "Agency Operations and Award Management," "National Science Board," and "Office of Inspector General." National Science Foundation, "FY 2018 Budget Request to Congress," https://www.nsf.gov/pubs/2017/nsf17057/nsf17057.pdf (accessed August 16, 2024).

25. The exception was RFP-III. In the summer of 2010, BP issued funding for time-sensitive data collection and monitoring in the immediate aftermath of the Deepwater Horizon spill. RFP-III offered an extension of that funding for up to two months. The initial round of funding, called GRI Year 1, predated the RFPs, hence the interruption in the numbering scheme posed by RFP-III.

26. Dr. John Farrington, notes on review of manuscript first draft, January 9, 2022.

27. Interview with Chuck Wilson, April 1, 2022.

28. Gulf of Mexico Research Initiative, *Request for Proposals (RFP-VI) for 2018–2019 for 1) Individual Investigators or Collaborative Efforts OR 2) Research Consortia*, October 3, 2016, https://gulfresearchinitiative.org/wp-content/uploads/2016/10/RFP-VI-Final.pdf (accessed August 16, 2024).

29. Interview with Leinen, March 24, 2021.

30. Comments from John Farrington, January 9, 2022.

31. Interview with Leinen, March 24, 2021.

32. Interview with Leinen, March 24, 2021.

33. American Institute of Biological Sciences, *AIBS: A Historical Perspective*, webpage, https://www.aibs.org/about-aibs/history.html (accessed August 16, 2024).

34. American Institute of Biological Sciences, *AIBS: A Historical Perspective*, webpage.

35. Conversation with Wilson, Shaw, and Carron, April 16, 2020.

36. Interview with Jennifer Pettit, May 26, 2020.

37. Interview with Jennifer Pettit, May 26, 2020.

38. Northern Gulf Institute, *Who We Are and What We Do*, webpage, https://www.northerngulfinstitute.org/about/who.php (accessed August 16, 2024). The six research institutions that participate in the Northern Gulf Institute are the University of Southern Mississippi, Louisiana State University, Florida State University, Dauphin Island Sea Lab, University of Alabama in Huntsville, and Mississippi State University, where the NGI is located.

39. Northern Gulf Institute, *Who We Are and What We Do*, webpage.

40. Interview with Jay Ritchie, July 23, 2020.

41. Consortium for Ocean Leadership, "Ocean Leadership Selects New Chair," webpage, June 25, 2008, https://oceanleadership.org/ocean-leadership-selects-new-chair/ (accessed August 27, 2021).

42. Consortium for Ocean Leadership, "Ocean Leadership Selects New Chair," webpage.

43. Interview with Katie Fillingham, February 5, 2019.

44. The summary can be found at Consortium for Ocean Leadership, *Deepwater Horizon Oil Spill: Scientific Symposium Meeting Summary*, Louisiana State University, June 3, 2010, https://oceanleadership.org/wp-content/uploads/2010/06/DeepwaterHorizonOilSpillSymposiumSummary.pdf (accessed August 27, 2021).

45. Interview with Steve Sempier, April 5, 2019.

46. Interview with Dr. Kenneth Halanych, April 27, 2020. Conversation with Wilson, Shaw, and Carron, April 16, 2020.

CHAPTER THREE: OPEN, TRANSPARENT, AND ACCESSIBLE DATA: GRIIDC AND GoMRI DATA MANAGEMENT POLICY

1. Edited from previous email to author, May 3, 2024.
2. Rossi, LeBel, and Gibeaut, 1.
3. Paraphrased from comments made by John Farrington to the author, June 7, 2022.
4. Correspondence with Chuck Wilson, June 11, 2024.
5. The Gulf of Mexico Research Initiative considers a consortium to be four or more institutions that work together to advance research about one or more of GoMRI's core areas. See Research Board, Gulf of Mexico Research Initiative, "Request for Proposals: Selection of Research Consortia," April 25, 2011, https://gulfresearchinitiative.org/wp-content/uploads/2011/03/GRI_RFP_I.pdf (accessed August 16, 2024).
6. "Request for Proposals: Selection of Research Consortia."
7. "Request for Proposals: Selection of Research Consortia."
8. "Request for Proposals: Selection of Research Consortia," 17.
9. "Request for Proposals: Selection of Research Consortia," 20.
10. Gulf of Mexico Research Initiative, "Request for Proposals for 2015–2017 GoMRI Research Consortia (RFP-IV)," November 15, 2013, https://gulfresearchinitiative.org/wp-content/uploads/2013/10/RFP-IV-Final.pdf (accessed August 16, 2024), 7.
11. "Request for Proposals for 2015–2017 GoMRI Research Consortia (RFP-IV)."
12. "Request for Proposals for 2015–2017 GoMRI Research Consortia (RFP-IV)," 12.
13. "Request for Proposals for 2015–2017 GoMRI Research Consortia (RFP-IV)."
14. Memorandum from the Gulf of Mexico Research Initiative Research Board to GoMRI (RFP-I) (RFP-II) Investigators, re: Compliance with Master Research Agreement Data Policies, December 2, 2013, https://gulfresearchinitiative.org/wp-content/uploads/2022/12/GoMRI_Data_Policy_Memo_2013.pdf (accessed August 16, 2024).
15. Memorandum from the Gulf of Mexico Research Initiative Research Board to GoMRI (RFP-I) (RFP-II) Investigators.
16. Memorandum from the Gulf of Mexico Research Initiative Research Board to GoMRI (RFP-I) (RFP-II) Investigators.

17. Rhiannon Meyers, "Edward H. Harte, Former Caller-Times Publisher and a Philanthropist, Dies at 88," *Corpus Christi Caller-Times*, May 18, 2011.

18. Harte Research Institute for Gulf of Mexico Studies, "Our History," https://www.harte.org/about (accessed August 16, 2024).

19. The professional information here was gathered from Dr. McKinney's CV, available at https://www.harteresearchinstitute.org/sites/default/files/2019-07/MCKINNEY%20RESUME%20GENERAL%20SHORT%20%287-16-19%29.pdf (accessed August 16, 2024).

20. The professional information presented here was gathered from both an interview with Dr. Gibeuat as well as his curriculum vitae, available at https://www.harteresearchinstitute.org/sites/default/files/2016-10/gibeaut_cv.pdf (accessed August 16, 2024).

21. Interview with Dr. James Gibeaut, December 6, 2019.

22. Harte Research Institute for Gulf of Mexico Studies, webpage, "People," "James Gibeaut, Ph.D. Endowed Chair for Coastal and Marine Geospatial Sciences," https://www.harteresearchinstitute.org/people/james-jim-gibeaut (accessed August 16, 2024).

23. https://www.harteresearchinstitute.org/sites/default/files/2016-10/gibeaut_cv.pdf (accessed August 16, 2024).

24. Interview with Dr. Gibeaut, December 6, 2019. The collaborative nature of GoMRI's staff is something that many participants in the initiative remember strongly. GoMRI had a bare-bones administrative and managerial staff, and all of its funding was disbursed through another organization (GOMA). Partner organizations like Ocean Leadership and HRI were necessary to the proper functioning of the initiative. A strong culture of camaraderie developed among the different people participating in GoMRI efforts, even though many worked for different organizations.

25. Interview with Dr. Gibeaut, December 6, 2019.

26. James Gibeaut, "Enabling Data Sharing through the Gulf of Mexico Research Initiative Information and Data Cooperative," 35.

27. Rossi, LeBel, and Gibeaut, 2.

28. Rossi, LeBel, and Gibeaut, 5.

29. The subsite can be found at gomri.griidc.org (accessed August 13, 2024).

30. Gulf of Mexico Research Initiative, "Master Research Agreement between BP Exploration and Production, Inc., and Gulf of Mexico Alliance," amended and restated December 1, 2011.

31. Interview with Dr. Gibeaut, December 6, 2019.

32. Information on the quality and quantity of data available in the wake of Ixtoc I and *Exxon Valdez* acquired from interview with Dr. Gibeaut.

33. Interview with Dr. Gibeaut, December 6, 2019.

34. Interview with Dr. Gibeaut, December 6, 2019.

35. Memorandum from Gulf of Mexico Research Initiative Research Board to GoMRI (RFP-I) (RFP-II) Investigators, "Compliance with Master Research Agreement Data Policies."

36. Memorandum from Rita Colwell, Chair, Gulf of Mexico Research Initiative Research Board to GoMRI (RFP-I) (RFP-II) (RFP-IV) Investigators, "Clarification and update of GOMRI Data Policy," https://gulfresearchinitiative.org/wp-content/uploads/2022/12/Memorandum-from-GOMRI-RB-Chair-Data-Policy-9-4-15.pdf (accessed August 16, 2024).

37. Memorandum from Rita Colwell, "Clarification and update of GOMRI Data Policy."

38. Gulf of Mexico Research Initiative, "Request for Proposals (RFP-VI) for 2018–2019 for 1) Individual Investigators or Collaborative Efforts OR 2) Research Consortia," https://gulfresearchinitiative.org/wp-content/uploads/2016/10/RFP-VI-Final.pdf (accessed August 16, 2024).

39. Memorandum from Chuck Wilson to Gulf of Mexico Research Initiative (GoMRI) RFP-VI Research Consortia and Individual Investigators, regarding "Update of GoMRI Data Policy and Procedures," August 29, 2017, https://gulfresearchinitiative.org/wp-content/uploads/2017/09/Update-of-GoMRI-Data-Policy-and-Procedures-Memorandum-August-29-2017.pdf (accessed August 16, 2024).

40. Memorandum from Chuck Wilson, "Update of GoMRI Data Policy and Procedures."

41. Interview with Dr. Gibeaut, December 6, 2019; interview with Kevin Shaw, October 10, 2019.

42. Gulf of Mexico Research Initiative, "Gulf of Mexico Research Initiative Annual Report (Year 8—2017)," 30.

43. Zimmermann, "From Disaster to Understanding: Formation and Accomplishments of the Gulf of Mexico Research Initiative," 22.

44. Rossi, LeBel, and Gibeaut, 1.

45. Rossi, LeBel, and Gibeaut, 6.

46. Email from Rosalie Rossi to Chuck Wilson, September 18, 2024. Provided to the author by Chuck Wilson.

CHAPTER FOUR: OUTREACH AND CAPACITY: BUILDING A LEGACY OF OIL SPILL SCIENCE IN THE GULF OF MEXICO

1. Larry McKinney, "America's Sea: We All Have a Stake in the Future of the Gulf of Mexico," *Texas Parks and Wildlife Magazine*, July 2011, https://tpwmagazine.com/archive/2011/jul/ed_2_gulf/ (accessed August 16, 2024). McKinney played a key role in the formation of GoMRI, especially pertaining to the organization's data management.

2. McKinney, "America's Sea: We All Have a Stake in the Future of the Gulf of Mexico."

3. Gulf of Mexico Alliance Water Quality Team, Monitoring Workgroup, White Paper Writing Team, "White Paper on Gulf of Mexico Water-Quality Monitoring: Providing Water-Quality Information to Support Informed Resource Management and Public Knowledge," December 2013, 7, https://sarasota.wateratlas.usf.edu/upload/documents/GOMA-Gulf-monitoring-white-paper-Final.pdf (accessed August 16, 2024).

4. "White Paper on Gulf of Mexico Water-Quality Monitoring."

5. Larry McKinney, "How Deepwater Horizon Changed the Trajectory of Gulf Research (Some Winners, Some Losers, and a Few Still Finding Their Footing," delivered during the 2020 GoMOSES Opening Plenary, February 4, 2020.

6. By focusing on C-IMAGE, I provide an example of how GoMRI-funded consortia developed outreach programming. I do not mean to imply that other consortia did not undertake exceptional outreach programming; citing all of the other consortia-led outreach efforts would take far more space than practical. Readers can find a list of the consortia in Appendix C. As of 2024, many still have operational websites, where one can learn more about their outreach programs.

7. C-IMAGE III, "Who We Are," https://www.marine.usf.edu/c-image/who-we-are/ (accessed August 16, 2024).

8. Sadly, Dr. Hollander passed away in October 2020, prior to the close-out of C-IMAGE. He had been involved with GoMRI-funded research since 2011. USF College of Marine Science, webpage, "Remembering David Hollander, PhD (1959–2020)," https://www.usf.edu/marine-science/news/2020/david-j-hollander-phd-1959-2020.aspx (accessed August 16, 2024).

9. C-IMAGE III, webpage, "Look and Listen," https://www.marine.usf.edu/c-image/media-player/ (accessed August 16, 2024).

10. C-IMAGE, webpage, "Teacher @ Sea," https://www.marine.usf.edu/c-image/teacher-sea/ (accessed August 16, 2024).

11. C-IMAGE, webpage, "Teacher @ Sea."

12. Debra S. Benoit, Leigh A. Zimmermann, et al., "An Opportunity to Inform and Educate through the Gulf of Mexico Research Initiative: Outreach Efforts Surrounding the Deepwater Horizon Oil Spill," *Oceanography* 29, no. 3 (September 2016): 38.

13. Benoit, Zimmermann, et al., 38–39.

14. Benoit, Zimmermann, et al.

15. Sea Grant, webpage, "About."

16. Benoit, Zimmerman, et al., 39.

17. Benoit, Zimmerman, et al., 40.

18. Benoit, Zimmerman, et al., 42.

19. Benoit, Zimmerman, et al., 43.

20. In the event that any of the links to Sea Grant literature no longer provide access to the cited material, they can be found at the National Sea Grant Library in Narragansett, Rhode Island. The library can be searched online here: https://repository.library.noaa.gov/cbrowse?pid=noaa%3A11&parentId=noaa%3A11 (accessed September 18, 2024).

21. Zimmerman, "From Disaster to Understanding: Formation and Accomplishments of the Gulf of Mexico Research Initiative," 25.

22. Benoit, Zimmerman, et al., 43.

23. Benoit, Zimmerman, et al., 43. In 2020, the Smithsonian Ocean Portal had 32,000 followers on Twitter and 128,000 followers on Facebook. Its social media influence amplified and broadcasted GoMRI research to a wide variety of stakeholders, most prominent among them being the public.

24. Zimmermann, "From Disaster to Understanding: Formation and Accomplishments of the Gulf of Mexico Research Initiative," 25.

25. Benoit, Zimmerman, et al., 44.

26. Benoit, Zimmerman, et al.

27. Benoit, Zimmerman, et al.

28. Benoit, Zimmerman, et al.

29. Most GoMRI-related video content is available here: https://www.youtube.com/c/GulfofMexicoResearchInitiative/videos (accessed September 18, 2024).

30. Gulf Coast Ecosystem Restore Council, *2018 Annual Report to Congress*, March 2019, 6, https://www.restorethegulf.gov/sites/default/files/2018_Annual_Report_to_Congress_508Compliant_20190306.pdf (accessed August 3, 2020). For a more precise breakdown of how the Gulf Coast Ecosystem Restoration Trust Fund was spent, see Gulf Coast Ecosystem Restoration Council, *Restoration Council: 10-Year Commemoration Report*, April 2020, https://www.restorethegulf.gov/sites/default/files/RestoreCouncil_10YearReport2020_v6.pdf (accessed April 29, 2022), 7. The report states that funding was divided into five buckets comprising different proportions of the 80 percent of Clean Water Act penalties. They were the Direct Component (35 percent) that was divided equally among the Gulf states "for ecosystem restoration and economic development"; the Council-Selected Restoration Component (30 percent plus 50 percent of the interest earned by the federal government's oil spill trust fund, into which the remaining 20 percent of the Clean Water Act penalties went) which was administered by the council for ecosystem restoration; the Spill Impact Component (30 percent), which was divided among the five states based on the proportion of Gulf of Mexico coastline within their jurisdiction that experienced oiling for relevant state projects and programs approved by the Council; the Gulf Coast Ecosystem Restoration Science Program (2.5 percent plus 25 percent of the interest earned by the trust fund) "for monitoring, observation, science and technology program administered by NOAA; and the Centers of Excellence Research Grants Program

(2.5% plus 25% of the interest earned by the trust fund), administered by the U.S. Treasury for a Center of Excellence in each Gulf Coast state." Relatively little of this funding went to scientific research, which strengthened the case for a scientific funding organization on the scale of GoMRI.

31. "GoMRI Master Research Agreement as Amended and Restated December 1st, 2011," 19.

32. Dashboard, webpage, GoMRI.org, https://gomri.org/admin-dashboard/.

33. Interview with Colwell, March 25, 2021.

34. Comments from John Farrington, June 7, 2022.

CHAPTER FIVE: GOMRI REFLECTS: SYNTHESIS AND LEGACY

1. Email to author, May 13, 2024.

2. Carina Wyborn, Elena Louder, et al., "Understanding the Impacts of Research Synthesis," *Environmental Science and Policy* 86 (2018): 72.

3. Joel Achenbach, "Why Do Many Reasonable People Doubt Science?" *National Geographic*, https://oupub.etsu.edu/uschool/faculty/tadlockd/documents/why_doubt_science.pdf (accessed August 16, 2024).

4. Gulf of Mexico Research Initiative, "GoMRI Synthesis and Legacy," webpage, https://gulfresearchinitiative.org/gomri-synthesis/ (accessed August 16, 2024).

5. Gulf of Mexico Research Initiative, webpage, "GoMRI Synthesis and Legacy."

6. Interview with Michael Feldman, February 5, 2020.

7. Gulf of Mexico Research Initiative, "Core Area 8—Knowledge Exchange with User Communities," https://gulfresearchinitiative.org/wp-content/uploads/2021/11/Core-Area-8-Description_Members.pdf (accessed August 16, 2024).

8. Gulf of Mexico Research Initiative, "Core Area 8—Knowledge Exchange with User Communities."

9. Gulf of Mexico Research Initiative, "Core Area 8—Knowledge Exchange with User Communities."

10. GoMRI uses the term *products* to refer to the various documents and visual materials that emerge from each core area's synthesis efforts. Products range from journal articles and book chapters to workshop proceedings and press releases. See https://gulfresearchinitiative.org/gomri-synthesis/products/.

11. Gulf of Mexico Research Initiative, "GoMRI Synthesis and Legacy," webpage.

12. Gulf of Mexico Research Initiative, "GoMRI Synthesis and Legacy," webpage.

13. RFP-VI, https://gulfresearchinitiative.org/wp-content/uploads/RFP-VI-Final.pdf (accessed August 31, 2021).

14. Email from Wilson to author.

15. Gulf of Mexico Research Initiative, Newsletter, "Capping Off the Decade of GoMRI," ttps://gulfresearchinitiative.org/wp-content/uploads/2021/07/Farewell 2020_GoMRI_Newsletter_FINAL-1.pdf (accessed August 16, 2024).

16. Gulf of Mexico Research Initiative, "Synthesis and Legacy Frequently Asked Questions," webpage.

17. Gulf of Mexico Research Initiative, "Synthesis and Legacy Frequently Asked Questions," webpage.

18. See *Oceanography: The Official Magazine of the Oceanography Society* 34, no. 1 (March 2021): 15–239.

19. Gulf of Mexico Research Initiative, "Synthesis and Legacy Frequently Asked Questions," webpage.

20. Michel Boufadel et al., "Physical Transport Processes that Affect the Distribution of Oil in the Gulf of Mexico: Observations and Modeling," *Oceanography* 34, no. 1 (March 2021); 59.

21. "What Is Synthetic Aperture Radar?," NASA webpage, https://www.earthdata.nasa.gov/learn/backgrounders/what-is-sar

22. "What Is Synthetic Aperture Radar?," NASA webpage.

23. Boufadel et al., 59.

24. Boufadel et al., 59.

25. Boufadel et al., 59.

26. Boufadel et al., 59.

27. Guillaume Novelli et al., "A Biodegradable Surface Drifter for Ocean Sampling on a Massive Scale," *Journal of Atmospheric and Oceanic Technology* 34, no. 11 (November 2017): 2510.

28. Nilde Maggie Dannreuther, David Halpern, Jürgen Rullkötter, and Dana Yoerger, "Technological Developments Since the Deepwater Horizon Oil Spill," *Oceanography* 34, no. 1 (March 2021): 197.

29. Novelli et al., 2510.

30. Dannreuther et al., 197.

31. Dannreuther et al., 196.

32. Dannreuther et al.

33. Dannreuther et al., 198.

34. See: Lakshmana D. Chandrala et al., "A Device for Measuring the *in-situ* Response of Human Bronchial Epithelial Cells to Airborne Environmental Agents," *Scientific Reports* 9, 7263 (2019): 1.

35. Dannreuther et al., 203.

36. Dannreuther et al.

37. Boufadel, 59.

38. John W. Farrington, Edward B, Overton, and Uta Passow, "Biogeochemical Processes Affecting the Fate of Discharged Deepwater Horizon Gas and Oil: New Insights and Remaining Gaps in Our Understanding," *Oceanography* 34, no. 1 (March 2021): 85.

39. C-IMAGE III, "MOSSFA: Marine Oil Snow Sedimentation and Flocculent Accumulation," https://www.marine.usf.edu/c-image/marine-snow-working-group/ (accessed July 29, 2024).

40. "MOSSFA: Marine Oil Snow Sedimentation and Flocculent Accumulation."

41. Farrington, Overton, and Passow, 85.

42. Kenneth M. Halanych et al., "Effects of Petroleum By-Products and Dispersants on Ecosystems," *Oceanography* 34, no. 1 (March 2021): 157.

43. Wilson et al., 231.

44. See, for example, Ari Kelman, "Boundary Issues: Clarifying New Orleans's Murky Edges," *Journal of American History* 94, no. 3 (December 2007): 695–703; Andy Horowitz, *Katrina: A History, 1915–2015* (Cambridge, MA: Harvard University Press, 2020).

45. David G. Westerholm et al., "Preparedness, Planning, and Advances in Operational Response," *Oceanography* 34, no. 1 (March 2021): 222.

46. Westerholm et al., 222–23.

47. Antoinetta Quigg et al., "A Decade of GoMRI Dispersant Science: Lessons Learned and Recommendations for the Future," *Oceanography* 34, no. 1 (March 2021): 100.

48. Quigg et al., 99.

49. Quigg et al., 105–6.

50. Quigg et al., 107.

51. Shannon Wieman et al., "GoMRI Insights into Microbial Genetics and Hydrocarbon Bioremediation Response in Marine Ecosystems," *Oceanography* 34, no. 1 (March 2021): 125.

52. Wieman et al., 126.

53. Wieman et al., 131.

54. Wieman et al.

55. Wieman et al.

56. Wilson et al., 236.

57. Paul A. Sandifer et al., "Human Health and Socioeconomic Effects of the Deepwater Horizon Oil Spill in the Gulf of Mexico," *Oceanography* 34, no. 1 (March 2021): 184.

58. Sandifer et al., 185.

59. Sandifer et al., 186.

60. Gulf of Mexico Research Initiative, webpage, "Synthesis and Legacy Frequently Asked Questions."

CONCLUSION: THE GOMRI MODEL: SOCIAL CONSIDERATIONS AND THE BEST SCIENCE

1. With 813 total articles, by far the most researched of the five themes was the "environmental effects of the petroleum/dispersant system on the sea floor, water column, coastal waters, beach sediments, wetlands, marshes, and organisms; and the science of ecosystem recovery." Theme five, or the "impact of oil spills on public health including behavioral, socioeconomic, environmental risk assessment, community capacity and other population health considerations and issues" produced sixty-five articles, the fewest in number of the five theme areas. See the GoMRI.org dashboard, https://gomri.org/admin-dashboard/.

2. National Oceanic and Atmospheric Administration, webpage, "An Integrated Assessment of Oil and Gas Release into the Marine Environment at the Former Taylor Energy MC20 Site," https://repository.library.noaa.gov/view/noaa/20612 (accessed August 16, 2024). An undersea mudslide triggered by Hurricane Ivan in 2004 damaged a Taylor Energy company oil platform. Unlike Deepwater Horizon, responders and engineers were unable to cap the Taylor energy wellhead. To date, the damaged wellhead continues to release between 400 and 1,900 gallons of oil into the Gulf each day. In December 2021 Taylor Energy settled its longstanding suit with the federal government for $475 million and no claim of liability for the spill. See Sebastien Malo, "Taylor Energy, Feds reach $475 mln Settlement in Longest-Running Oil Spill," *Reuters*, December 22, 2021, https://www.reuters.com/legal/litigation/taylor-energy-feds-reach-475-mln-settlement-longest-running-oil-spill-2021-12-22/ (accessed August 16, 2024).

3. National Oceanic and Atmospheric Administration, webpage, "Oil Spills," https://www.noaa.gov/education/resource-collections/ocean-coasts/oil-spills (accessed August 16, 2024).

4. William R. Freudenberg and Robert Gramling, *Blowout in the Gulf: The BP Oil Spill Disaster and the Future of Energy in America* (Boston: MIT Press, 2011), 12–13.

5. Quoted in Andy Horowitz, *Katrina: A History* (Cambridge, MA: Harvard University Press, 2020), 193.

6. Horowitz, *Katrina: A History*, 190.

7. Comments from David R. Shaw to the author, June 6, 2022.

8. Comments from Singer, June 10, 2022.

9. Comments from Singer, June 10, 2022.

10. Interview with Chuck Wilson, April 1, 2022.

APPENDIX E: REQUEST FOR PROPOSAL TIMELINE

1. Based on information acquired from Leigh A. Zimmermann, Michael G. Feldman, Debra S. Benoit, Michael J. Carron, et al., "From Disaster to Understanding: Formation and Accomplishments of the Gulf of Mexico Research Initiative," *Oceanography* 34, no. 1 (March 2021): 16–29 (inclusive 16). Direct quotes describing the RFPs were taken from Gulf of Mexico Research Initiative, "GoMRI Funding Programs," website, https://research.gulfresearchinitiative.org/research-awards/ (accessed March 31, 2022).

INDEX

Allen, Thad, 17
American Institute of Biological Sciences (AIBS), 21, 26–27, 47, 65–66
American Petroleum Institute, 184n61
Army Corp of Engineers, 131

Barataria Bay, 131
Bay of Campeche, 4
Blanchard, Dean, 145–46
Boufadel, Michel, 126
Brewer, Peter, 16, 75, 79, 121
British Petroleum (BP), ix–x, 4, 6, 7, 14–15, 16–18, 19, 21–22, 23–24, 26, 27–28, 30, 31–33, 34, 35, 39, 40–42, 43, 45, 50, 53–55, 56, 57, 62, 64, 65–66, 67, 76–77, 79, 89–90, 95, 106, 107, 109, 143, 145–46, 151, 155, 182n35, 185n5, 185n8, 187n25

California Bay-Delta Authority, 25
Campbell, Robert, 37
Carron, Michael, 27, 42, 46, 52–53, 54, 55–56, 72
Center for the Integrated Modeling and Analysis of the Gulf Ecosystem (C-IMAGE), 96–98, 129, 148, 191n6
Chesapeake Bay, 94, 149
Chesapeake Bay Program, 25

Clean Water Act, 106–7, 192n30
Coast Guard, US, 14, 15, 17, 33, 40, 147
Colwell, Rita, 7, 8, 13, 19–20, 21–23, 28, 29, 30, 31, 32, 41, 43, 45, 49, 50, 53–54, 64, 65, 66, 72, 76, 77, 87, 91, 96, 108, 110, 120, 121–22, 151, 155, 186n8
Consortium for Advanced Research on Transport of Hydrocarbon in the Environment (CARTHE), 126–27, 129
Consortium for Ocean Leadership, 26, 47, 55, 58, 59, 60–61, 62, 65, 66, 68–70, 83, 99, 101, 144, 152, 189n24
Consortium for Oceanographic Research and Education (CORE), 69, 181n26
COVID-19, 73, 150, 180n6, 185n2

Damage Assessment, Remediation, and Restoration Program (DARRP), 16
Dannreuther, Maggie, 101
Deepwater Horizon oil spill, ix, 3–5, 7–8, 9, 10, 13–15, 17, 27, 32, 33–34, 35, 37, 38–39, 40–41, 43, 65, 67, 77, 82–83, 86, 88, 89–90, 93, 95, 97, 102, 103, 105, 106, 111, 116, 124, 125, 126, 128, 131, 134, 135, 136, 138–39, 140,

143, 145, 146–47, 148, 149, 150, 155, 187n25, 196n2
Delaware River Basin Commission, 25
Dispatches from the Gulf (Weiner), 105–6, 111, 149
dispersants, 29, 39–41, 117, 126, 129, 132–34, 135; and effects on marine ecosystem, 17, 29, 96, 115, 137, 138–39, 144, 196n1
Dudley, Robert, 15–16

Environmental Protection Agency (EPA), 16, 25, 138, 182n26
Environmental Science and Policy, 25
Exxon Valdez, 4, 10, 33, 38, 39, 42, 43, 83, 86, 133, 179n3, 180n8, 189n32

Farrington, John, 35, 36, 38, 46, 59, 129
Fleytas, Andrea, 14
Florida Department of Environmental Protection, 25
Florida Institute of Oceanography (FIO), 51

Gibeaut, James, 83–84, 85, 87, 88, 89, 90
Grand Lagrangian Deployment (GLAD), 127
Great Lakes, 20, 24, 25, 70, 94, 100, 149
Gulf Coast Ecosystem Restoration Trust Fund, 106, 192n30
Gulf of Mexico Alliance, 24–28, 30, 55, 65–66, 94, 107, 109; and conflicts of interest, 27–28; origins of, 24–25
Gulf of Mexico Oil Spill and Ecosystem Science Conference (GoMOSES), 54, 83, 94, 105–6, 109–10, 111, 146, 150
Gulf of Mexico Program, 25
Gulf of Mexico Research Initiative (GoMRI): and Data Committee, 78, 186n15; and data management, 8–9, 31, 33, 39, 52, 62–63, 76–91, 148; and equitable allocation of funding, 28–29; and funding, 4, 9, 6, 10, 11, 19, 23, 26–28, 30–31, 32–33, 36, 39, 42–43, 49, 50–51, 54, 55–57, 58–60, 62, 63–66, 70, 71, 73, 76–77, 79, 81, 84–85, 87–88, 90, 91, 94–95, 97, 101, 106–9, 111, 115, 122, 125–26, 128–29, 132, 134–35, 139–40, 144–45, 148, 150–55; and independence from BP, 32; and management structure, 46–49; and Master Research Agreement, 27–28, 29, 32, 45, 47, 49–50, 51, 52, 54–55, 56, 64, 76, 77, 78, 80, 81–82, 85, 86, 87, 90, 96, 107, 109, 118, 137, 143, 185n8; and oil spill science, 5–6, 9, 18, 70, 104–5, 107–11, 115, 116, 121, 122, 148–53; and outreach, x, 5, 6, 7, 8, 9, 10, 20, 46, 63, 67, 68–70, 93–111; and partnerships, 64–65; and request for proposals (RFP), 22, 32, 51, 57–58, 59–63, 68, 79–81, 87–88, 96–97, 98, 99, 121–22, 140, 144, 148–49, 153; and Synthesis and Legacy Committee, 116, 118, 119, 121, 123, 124, 151, 152
Gulf of Mexico Research Initiative Information and Data Cooperative (GRIIDC), x, 8, 9, 33, 48, 75–77, 79, 80–92
Gulf of Mexico Research Initiative Research Information System (RIS), 67–68
Gulf of Mexico Sea Grant Oil Spill Science and Outreach Program (SGOSSP), 102–3
Gulf of Mexico Sea Grant Oil Spill Science Outreach Team, 70–71
Gulf of Mexico University Research Collaborative, 82

Index

Halpern, David, 22
Harte, Ed, 82
Harte Research Institute (HRI), 25, 48, 54, 81, 82, 83, 84, 93
Hayward, Tony, 15, 41
Hogarth, William T., 51
Holdren, John, 16–17
Hollander, David, 97, 191n8
Horowitz, Andy, 147
Hurricane Harvey, 48, 89, 136
Hurricane Ivan, 196n2
Hurricane Katrina, 102, 136

Imberger, Jorg, 22
Ixtoc I, 4, 34–37, 42, 43, 86, 97; lack of scientific study of, 36–37

Jackson, Lisa, 16, 41
Jindal, Bobby, 30
Johns Hopkins University Laboratory for Experimental Fluid Dynamics, 128
Joint Oceanographic Institutes (JOI), 69

LeBel, Deborah, 85, 90
Lehner, Peter, 42
Leinen, Margaret, 22, 28, 29, 45, 50, 93
Loop, The (podcast), 97
Louisiana State University, 18, 182n26, 187n38

Marine Oil Snow Sedimentation and Flocculent Accumulation (MOSSFA), 129–30
McKinney, Larry, 82–83, 93, 94, 150, 190n1
McNutt, Marcia, 17
Mississippi Department of Environmental Quality, 25
Mississippi State University, 67, 187n38

Monterey Bay Aquarium Research Institute, 18
Murawski, Steve, 36, 97

National Academy of Sciences, 42, 66
National Commission on the BP Deepwater Horizon Oil Spill and Offshore Drilling, 35
National Oceanic and Atmospheric Administration (NOAA), 15, 16, 23, 25, 32, 33, 35, 40, 67, 70, 84, 91, 100, 107, 119, 179, 182n26, 192n30, 196n2
National Oceanography Centre, 18, 22
National Resource Damage Assessment (NRDA), 126
National Science Board (NSB), 20, 186n24
National Science Foundation (NSF), 16, 19, 20–21, 32, 47, 50, 57
National Sea Grant, 10, 20, 32, 47, 60, 70–73, 100, 102–3, 104, 119, 146–47, 192n20
Natural Resources Defense Council, 41–42
North Atlantic Torrey Canyon oil spill, 38
Northern Gulf Institute (NGI), 27, 47, 55, 67–68, 187n38

Oceans Act of 2000, 24
Outreach and Communication Committee (OCC), 99

Petitt, Jennifer, 66
Petróleos Mexicanos (PEMEX), 34
Pew Research Center, 185n2

Quigg, Antonietta, 133, 134

Rossi, Rosalie, 85, 90
Rullkötter, Jürgen, 22

Salazar, Ken, 16
Sandifer, Paul A., 136–37, 182n26
Santa Barbara oil spill, 37
Screenscope Productions, 93, 102, 104–5, 106, 149
Scripps Institution of Oceanography, 18
Sempier, Steve, 71–72
Shaw, Kevin, 27, 46, 52, 53, 54–55, 56, 72, 87, 88
Shepherd, John, 119
Smithsonian Ocean Portal, 102, 103–4, 149, 192n23
Spezio, Teresa Sabol, 37
Spotts, Pete, 41, 42
Swann, LaDon, 70
Synthetic Aperture Radar (SAR), 126

Tampico Maru oil spill, 38
Taylor Energy oil spill, 37, 145, 196n2
Teacher @ Sea, 97–98
Texas A&M University–Corpus Christi, 18, 25, 90, 128

United States Commission on Ocean Policy, 24
United States Naval Oceanographic Office (NAVOCEANO), 55
University of Glasgow, 18
University of Maryland–College Park Sea Grant program, 28

Vinnem, Jan Erik, 36–37

Weiner, Hal and Marilyn, 105, 149
Williams, Ellen, 7, 14, 15, 17–18, 19, 21, 22, 49
Wilson, Charles "Chuck," 20, 30–31, 46, 50, 53, 54, 55, 56, 72, 84, 87–88, 90, 130, 135

Zimmermann, Leigh, 58

ABOUT THE AUTHOR

Justin Shapiro is postdoctoral associate in climate pedagogy at Duke University. He is a historian of the environment, technology, and climate. Justin received his PhD from the University of Maryland–College Park in 2020.

www.ingramcontent.com/pod-product-compliance
Lightning Source LLC
Chambersburg PA
CBHW030109170426
43198CB00009B/549